INITIATING THE
SAFETY INFRASTRUCTURE
FOR A NUCLEAR POWER PROGRAMME

The following States are Members of the International Atomic Energy Agency:

AFGHANISTAN
ALBANIA
ALGERIA
ANGOLA
ANTIGUA AND BARBUDA
ARGENTINA
ARMENIA
AUSTRALIA
AUSTRIA
AZERBAIJAN
BAHAMAS, THE
BAHRAIN
BANGLADESH
BARBADOS
BELARUS
BELGIUM
BELIZE
BENIN
BOLIVIA, PLURINATIONAL
 STATE OF
BOSNIA AND HERZEGOVINA
BOTSWANA
BRAZIL
BRUNEI DARUSSALAM
BULGARIA
BURKINA FASO
BURUNDI
CABO VERDE
CAMBODIA
CAMEROON
CANADA
CENTRAL AFRICAN
 REPUBLIC
CHAD
CHILE
CHINA
COLOMBIA
COMOROS
CONGO
COOK ISLANDS
COSTA RICA
CÔTE D'IVOIRE
CROATIA
CUBA
CYPRUS
CZECH REPUBLIC
DEMOCRATIC REPUBLIC
 OF THE CONGO
DENMARK
DJIBOUTI
DOMINICA
DOMINICAN REPUBLIC
ECUADOR
EGYPT
EL SALVADOR
ERITREA
ESTONIA
ESWATINI
ETHIOPIA
FIJI
FINLAND
FRANCE
GABON
GAMBIA, THE

GEORGIA
GERMANY
GHANA
GREECE
GRENADA
GUATEMALA
GUINEA
GUYANA
HAITI
HOLY SEE
HONDURAS
HUNGARY
ICELAND
INDIA
INDONESIA
IRAN, ISLAMIC REPUBLIC OF
IRAQ
IRELAND
ISRAEL
ITALY
JAMAICA
JAPAN
JORDAN
KAZAKHSTAN
KENYA
KOREA, REPUBLIC OF
KUWAIT
KYRGYZSTAN
LAO PEOPLE'S DEMOCRATIC
 REPUBLIC
LATVIA
LEBANON
LESOTHO
LIBERIA
LIBYA
LIECHTENSTEIN
LITHUANIA
LUXEMBOURG
MADAGASCAR
MALAWI
MALAYSIA
MALI
MALTA
MARSHALL ISLANDS
MAURITANIA
MAURITIUS
MEXICO
MONACO
MONGOLIA
MONTENEGRO
MOROCCO
MOZAMBIQUE
MYANMAR
NAMIBIA
NEPAL
NETHERLANDS,
 KINGDOM OF THE
NEW ZEALAND
NICARAGUA
NIGER
NIGERIA
NORTH MACEDONIA
NORWAY
OMAN

PAKISTAN
PALAU
PANAMA
PAPUA NEW GUINEA
PARAGUAY
PERU
PHILIPPINES
POLAND
PORTUGAL
QATAR
REPUBLIC OF MOLDOVA
ROMANIA
RUSSIAN FEDERATION
RWANDA
SAINT KITTS AND NEVIS
SAINT LUCIA
SAINT VINCENT AND
 THE GRENADINES
SAMOA
SAN MARINO
SAUDI ARABIA
SENEGAL
SERBIA
SEYCHELLES
SIERRA LEONE
SINGAPORE
SLOVAKIA
SLOVENIA
SOMALIA
SOUTH AFRICA
SPAIN
SRI LANKA
SUDAN
SWEDEN
SWITZERLAND
SYRIAN ARAB REPUBLIC
TAJIKISTAN
THAILAND
TOGO
TONGA
TRINIDAD AND TOBAGO
TUNISIA
TÜRKİYE
TURKMENISTAN
UGANDA
UKRAINE
UNITED ARAB EMIRATES
UNITED KINGDOM OF
 GREAT BRITAIN AND
 NORTHERN IRELAND
UNITED REPUBLIC OF TANZANIA
UNITED STATES OF AMERICA
URUGUAY
UZBEKISTAN
VANUATU
VENEZUELA, BOLIVARIAN
 REPUBLIC OF
VIET NAM
YEMEN
ZAMBIA
ZIMBABWE

The Agency's Statute was approved on 23 October 1956 by the Conference on the Statute of the IAEA held at United Nations Headquarters, New York; it entered into force on 29 July 1957. The Headquarters of the Agency are situated in Vienna. Its principal objective is "to accelerate and enlarge the contribution of atomic energy to peace, health and prosperity throughout the world".

IAEA-TECDOC-2109

INITIATING THE SAFETY INFRASTRUCTURE FOR A NUCLEAR POWER PROGRAMME

INTERNATIONAL ATOMIC ENERGY AGENCY
VIENNA, 2026

COPYRIGHT NOTICE

For further information on this publication, please contact:

Regulatory Activities Section
International Atomic Energy Agency
Vienna International Centre
PO Box 100
1400 Vienna, Austria
Email: Official.Mail@iaea.org

© IAEA, 2026
Printed by the IAEA in Austria
January 2026
https://doi.org/10.61092/iaea.u85k-q59c

IAEA Library Cataloguing in Publication Data

Names: International Atomic Energy Agency.
Title: Initiating the safety infrastructure for a nuclear power programme / International Atomic Energy Agency.
Description: Vienna : International Atomic Energy Agency, 2026. | Series: IAEA-TECDOC ISSN 1011-4289 ; no. 2109 | Includes bibliographical references.
Identifiers: IAEAL 26-01808 | ISBN 978-92-0-127525-7 (paperback : alk. paper) | ISBN 978-92-0-127625-4 (pdf)
Subjects: LCSH: Nuclear power plants — Safety measures. | Nuclear power plants — Management. | Nuclear power plants — International cooperation.

FOREWORD

The success of a national nuclear power programme depends on the strong commitment of the government to ensuring that nuclear power is used safely, securely and exclusively for peaceful purposes. This commitment includes establishing a strong, independent regulatory body to support the overall nuclear power programme in a timely manner.

The IAEA safety standards establish requirements and provide guidance to Member States on the role of the regulatory body and the nature of its work. This work evolves through the successive phases of a new nuclear power programme, namely the initial planning and strategy formulation; the establishment of the organization, recruitment, staffing and the development of the management system; fostering a strong culture for safety; the preparation of regulations, licensing and inspection of construction; assessment and decisions regarding the proposed arrangements for operation; and ultimately to the oversight of commissioning and operation of the nuclear power plant.

Each Member State has an approach to establishing its regulatory framework that depends on factors such as its existing status in relation to nuclear safety, the availability of resources, the planned technology and the size of the nuclear power programme.

This publication presents the experiences of several Member States that are in the process of establishing the safety infrastructure for their first nuclear power plant. Its objective is to provide information and practical examples of the main activities to be carried out during Phase 1 of the development of the nuclear safety infrastructure, by sharing the experiences of these States through case studies that describe the challenges faced, summarize the lessons learned and highlight good practices. The publication does not evaluate this information against IAEA safety standards or relevant guidance, since such evaluations are covered by the peer review and advisory services offered by the IAEA.

The IAEA is grateful to all the experts who contributed to the development and review of this publication, in particular H. Khouaja (Canada). The IAEA officers responsible for this publication were M. Aoki and R. Gomaa of the Division of Nuclear Installation Safety.

1. INTRODUCTION

1.1. BACKGROUND

The IAEA safety standards, in particular IAEA Standards Series Nos SF-1, Fundamental Safety Principles [1] and GSR Part 1 (Rev. 1), Governmental, Legal and Regulatory Framework for Safety [2], provide the foundations for countries initiating the development of a nuclear power programme. IAEA Safety Standards Series No. SSG-16 (Rev. 1), Establishing the Safety Infrastructure for a Nuclear Power Programme [3] provides recommendations for establishing the necessary safety infrastructure, presenting a holistic approach for the establishment of a national safety infrastructure that aligns with IAEA Nuclear Energy Series No. NG-G-3.1 (Rev. 1) Milestones in the Development of a National Infrastructure for Nuclear Power [4] (hereinafter referred to as the 'Milestones approach'. SSG-16 (Rev. 1) [3] provides 197 recommended actions to progressively meet all applicable IAEA safety requirements for developing the safety infrastructure necessary to support a nuclear power programme. These recommendations are organized into three phases as follows:

- Phase 1 (duration 1 to 3 years) is an initiation stage that includes considerations before the decision to launch a nuclear power programme is taken.

- Phase 2 (duration 3 to 7 years) is a preparatory stage for contracting and construction of a nuclear power plant after a policy decision has been taken.

- Phase 3 (duration 7 to 10 years) is an implementation stage in which activities covering contracting and construction of the first NPP, up to readiness for operation.

To assist embarking countries with little or no prior experience in implementing a nuclear power plant (NPP), the IAEA developed a Generic Road Map (GRM) to inform on the major steps to establish and maintain a comprehensive safety infrastructure for licensing a first NPP [5]. The GRM is supported by a series of topical publications to supplement existing IAEA safety standards. The publications are intended to incorporate lessons learned from embarking countries that have advanced in the development of their nuclear power programmes or have recently started commercial operation of NPPs, as well as from countries that are currently constructing or commissioning their first NPP unit(s). This TECDOC is one of these publications.

This TECDOC focuses on the experience, lessons learned, and best practices of embarking countries in Phase 1. In line with the recommendations of SSG-16 (Rev. 1) [3], six embarking countries, at different stages of nuclear power programme development, with different nuclear histories, resources and environments, have provided case studies on initiating a nuclear power programme.

1.2. OBJECTIVE

The objective of this publication is to provide the governments of countries embarking on a nuclear power programme with information and practical examples of the main activities to be carried out during Phase 1 of the development of the nuclear safety infrastructure, to support the government in making an informed decision on whether to launch a nuclear power programme. In Phase 1, prior to the decision to launch a nuclear power programme, para. 2.16 of SSG-16 (Rev. 1) [3] recommends that:

"Before making a knowledgeable decision with regard to the introduction of a nuclear power programme, the government should ensure that all expected impacts of the decision are thoroughly understood and that an adequate assessment of the State's safety infrastructure and associated needs has been conducted. At the end of Phase 1, the government should be fully aware that embarking on a nuclear power programme involves a firm and long-term commitment to maintaining activities that are necessary for ensuring safety".

1.3. SCOPE

This TECDOC provides an analysis of the experiences of the six Member States that participated in a case study to share their experiences and lessons learned in developing the safety infrastructure necessary to launch or support a new nuclear power programme.

Belarus, Egypt, Ghana, Nigeria, United Arab Emirates (UAE) and Uzbekistan participated in the development of this TECDOC. These Member States, whose nuclear programmes are at different stages of development, provided their experiences on challenges encountered, solutions implemented, and good practices during Phase 1.

A series of topics related to safety infrastructure are presented for discussion. Each topic begins with a brief introduction of the associated issues, referencing relevant IAEA safety standards and additional useful resources. This leads to the identification of key points of interest, followed by an analysis of trends and notable aspects in the experience of each country based on the information provided in the case study reports.

A range of safety infrastructure topics are introduced for discussion. Each topic starts with a concise overview of the related issues, citing pertinent IAEA safety standards and other valuable resources. This is followed by highlighting key points of interest and analysing trends and significant aspects from each country's experiences as detailed in the case study reports.

This TECDOC does not cover the full set of Phase 1 actions for all related safety elements, as delineated in SSG-16 (Rev. 1) [3]. To that end, a review of IAEA missions[1] [6] [7] conducted in embarking countries was completed and is reflected in the analysis part of the publication. The selection of the key safety elements addressed herein reflects those areas that Member States have experienced the most issues in implementing, based on the number of recommendations and suggestions made in the mission reports. Infrastructure actions associated with nuclear security or safeguards are addressed in other IAEA publications (e.g. [8][9]).

1.4. STRUCTURE

The publication is divided into four sections and six annexes.

Section 2 presents the current status of the nuclear power programmes in the case study countries as per July 2023.

Section 3 presents experience of embarking countries in Phase 1 in relation to their legal, governmental, and regulatory frameworks. Details of shared and/or unique practices and outcomes are described. Challenges encountered, solutions implemented, and good practices are presented.

---

[1] Integrated Regulatory Review Service (IRRS) and Integrated Nuclear Infrastructure Review (INIR) missions.

Similarly, Section 4 presents key areas of safety infrastructure that need to be well understood prior to introducing a nuclear power programme; namely, a general site survey, consultation with interested parties, establishment of leadership and management for safety, and human resources development.

The six Annexes contain case study reports from the Member States involved. The information in these reports has been provided by country experts and its inclusion in this publication does not imply the IAEA's endorsement. The IAEA cannot guarantee the accuracy or completeness of this information. The main text of the TECDOC does not evaluate the annexes' information against IAEA safety standards, security recommendations, or relevant guidance. Instead, it provides a summary of common challenges and issues, highlights implemented solutions and good practices and references the outcomes of IAEA peer review and advisory services conducted in those countries.

Additional information on each country's practices is available on the IAEA's web page on Country Nuclear Power Profiles [10] national reports to the Convention on Nuclear Safety (CNS) [11] Joint Convention on the Safety of Spent Fuel Management and on the Safety of Radioactive Waste Management [12], and in peer review and advisory mission reports available on the IAEA's website.

2. CURRENT STATUS OF THE NUCLEAR POWER PROGRAMME IN CASE STUDY COUNTRIES

The nuclear power programmes at the six countries that participated in the case studies continued to evolve, while the TECDOC presents a snapshot of nuclear power programmes at different levels of maturity at a given point in time (July 2023) which reflects the freezing date for the case study reports. In two countries — Ghana and Nigeria — the governments have indicated their intention to implement a nuclear power programme but, as of July 2023, no concrete steps have been taken toward NPP construction. In Uzbekistan, several agreements and Memorandums of Understanding (MOUs) have been signed between the government of Uzbekistan and the Russian Federation. In, Egypt, contracts are in place for the supply and construction of NPPs at the designated site, and the first concrete pouring that took place in July 2022 represented a milestone. The remaining two countries — Belarus and the UAE — have essentially completed the construction and commissioning phases and have put their NPPs into operation. Figure 1 presents the current status of the case study countries relative to the IAEA Milestones phases [4].

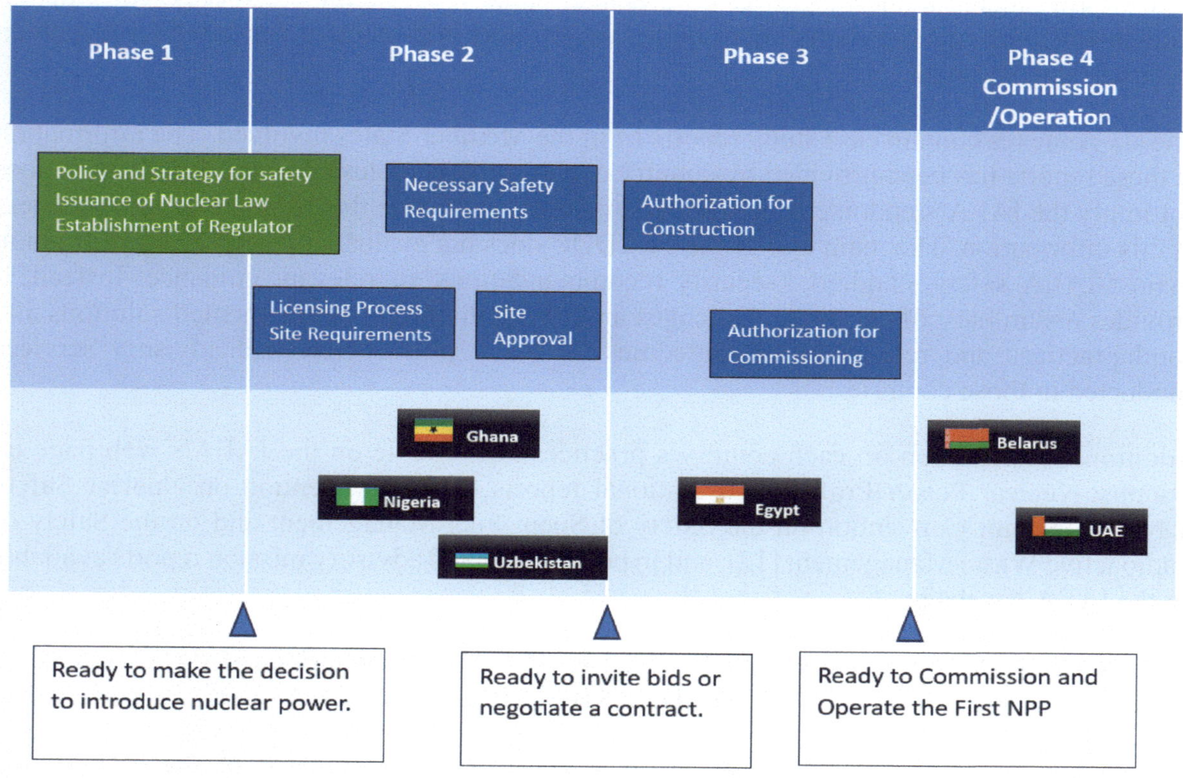

FIG. 1. The status of the case study countries' programmes relative to the IAEA Milestones phases

The status of the nuclear power programme in each country is summarized below, countries were introduced in alphabetical order.

2.1. BELARUS

In 2011, the governments of Belarus and the Russian Federation signed an agreement for cooperation to construct an NPP in Belarus. The Belarusian NPP located in Ostrovets, consists of two AES-2006 design VVER (water cooled, water moderated power reactor) units with a total capacity of 2400 MWe. The construction was performed by Atomstroyexport on a turnkey basis, and 'Belarusian NPP' serves as the customer and operator. A licence for the full range of construction activities for Unit 1 of the Belarusian NPP was issued in April 2014 and for Unit 2 in December 2014. Unit 1 entered into operation in 2020 and Unit 2 in 2023 [13]. A similar NPP design is utilized in the Leningrad-2 NPP project in the Russian Federation, which has been in operation since March 2018.

2.2. EGYPT

In November 2015, Egypt and the Russian Federation concluded an intergovernmental agreement to build and operate four reactors at the El Dabaa site, including fuel supply, used fuel, training, and development of regulatory infrastructure. In April 2019, the Nuclear Power Plants Authority (NPPA) received a site approval permit for the El Dabaa site from the Egyptian Nuclear and Radiological Regulatory Authority (ENRRA) [14]. The planned units at El Dabaa are similar to those operating at Leningrad-2. In June 2022, ENRRA approved the construction permit for Unit 1. Construction for Unit 2 began in November 2022, and in March 2023, ENRRA issued a construction permit for Unit 3. First concrete pouring of Unit 1 in July 2022 was marked as a milestone. All four units are under construction in 2025.

2.3. GHANA

Ghana operates a research reactor (Ghana Research Reactor-1), which started operation in March 1995. Ghana has also declared its intention to launch a nuclear power programme and has been working to complete the planning for its implementation. A Programme Comprehensive Report was submitted to the Government to support a decision on the nuclear power programme, and an Inter-Ministerial Committee was constituted to address funding options and interact with neighbouring countries on the introduction of nuclear power in Ghana. The Government announced the decision to embark on nuclear power in August 2022. Large reactors and small modular reactors are under consideration by Nuclear Power Ghana. Ghana enacted the Nuclear Regulatory Authority Act (Act 895) in 2015 establishing the Nuclear Regulatory Authority (NRA) and has been working to develop the necessary safety infrastructure [15].

2.4. NIGERIA

Nigeria currently operates a multi-purpose research reactor and has declared its intention to embark on a nuclear power programme. To prepare for the implementation of such a programme, Nigeria is currently developing infrastructure, including building national institutions, establishing a legal and regulatory framework, developing human resources and financial strategies, and addressing radioactive waste management while involving interested parties. These activities will facilitate the construction of the first nuclear power plant by adopting the IAEA Milestones approach. This work is coordinated through the Nuclear Energy Programme Implementation Committee (NEPIC) [24].

2.5. UNITED ARAB EMIRATES

In April 2008, the UAE government released its 'Policy of the United Arab Emirates on the Evaluation and Potential Development of Peaceful Nuclear Energy'[16] which set out the rationale and goals for a proposed nuclear power programme. Thereafter, in December 2009, the Emirates Nuclear Energy Company (ENEC) announced that it had selected a consortium led by the Korea Electric Power Corporation (KEPCO) to design, build and assist in the operation and maintenance of four 1400 MWe APR-1400 units. A construction licence was issued in 2012 for the first two units at the Barakah site located on the coast west of Abu Dhabi, followed in 2014 by the construction licence for Units 3 and 4. Units 1 to 3 at Barakah are now in commercial operation and connected to the grid. A similar NPP design is utilized in Saeul 1 and 2 (previously Shin Kori 3 and 4) in the Republic of Korea, which have been in operation since 2016 and 2019, respectively.

2.6. UZBEKISTAN

In September 2018, an intergovernmental agreement was signed between Uzbekistan and the Russian Federation for the construction by Rosatom of two VVER-1200 reactors to be commissioned around 2028. The reference units are Novovoronezh II. In addition, on July 19, 2018, the Decree of the President of the Republic of Uzbekistan No. 5484 "On measures for the development of nuclear energy in the Republic of Uzbekistan" [17] was approved.

3. ROLES AND RESPONSIBILITIES OF THE GOVERNMENT AND THE REGULATORY BODY IN PHASE 1 OF THE NUCLEAR POWER PROGRAMME

3.1. NATIONAL POLICY AND STRATEGY FOR SAFETY

3.1.1. The issue

A nuclear power programme is a major national initiative that needs careful planning and preparation, along with significant investments of time, human and financial resources. Considering the wide spectrum of issues to be considered and the implications and duration of the commitments associated with a nuclear power programme, the decision to embark on a such a programme has to come from the government. The importance of safety has to be recognized and reflected in the policy decisions as well as in the strategy adopted by the government.

3.1.2. Points of interest

How the national governments in the case study countries implemented the actions recommended by SSG-16 (Rev. 1) [3] in Phase 1 regarding the national policy and strategy for safety, and what challenges and solutions they experienced, namely:

> **"Action 1. The government should consider the necessary elements of a national policy and strategy for safety to meet the fundamental safety objective and safety principles established in SF-1 [1].**

> **"Action 2. The government should provide for the coordination of all activities to establish the safety infrastructure.**

> **"Action 3. The government should ensure that the status of the safety infrastructure in relevant areas is assessed and that radiation protection considerations are adequately taken into account.**

> **"Action 4. The government should take due account of the assessment of the elements of the safety infrastructure and the fundamental principle of justification when making a decision on whether to introduce a nuclear power programme."**

3.1.3. Discussion

The national government of each case study country has issued, or has at least drafted, policy statements on the adoption of nuclear in the national energy plans. In several cases, for instance in Egypt, Ghana, and Nigeria, several years elapsed between the first issuance of the government policy statement on the use of nuclear power and the actual implementation of a programme, reflecting the long-standing ambition of these countries to develop nuclear power. In some cases, the national position went through several iterations for reasons including changing of the national circumstances, the need for wide interested parties' consultation, and to conform with IAEA safety standards. In other countries, notably the UAE and Uzbekistan, the issuance of a policy on nuclear energy was followed quickly by action to initiate the nuclear programme.

In all countries studied, the national government plays a central role in coordinating the actions to establish the needed safety infrastructure, typically through the establishment of a designated nuclear energy programme implementing organization (NEPIO) and/or a high-level coordinating committee composed of representatives from the relevant ministries, state agencies, and technical staff to oversee the development of infrastructure across different areas.

However, feedback from various Integrated Regulatory Review Service (IRRS) missions highlights the challenges related to full correspondence with the national policy and strategy for safety as specified by Requirement 1 of GSR Part 1 (Rev. 1) [2]. The UAE practice comes closest to this standard with its Policy on the Evaluation and Potential Development of Peaceful Nuclear Energy that the Cabinet of Ministers issued before the launch of the UAE nuclear programme. An IRRS review mission to the UAE observed that such policies are rarely specifically formulated even in countries with mature nuclear power programmes. The IRRS mission considered that the Nuclear Policy and the subsequent Nuclear Law constituted an adequate platform for addressing the Fundamental Safety Principles [1] and identified it as a good practice.

Principle 4 of SF-1 states that "**Facilities and activities that give rise to radiation risks must yield an overall benefit.**" Formal risk-benefit justifications for the proposed nuclear power programmes were generally not explicitly stated in the pre-feasibility study reports, although the study reports describe both risk and benefit, and the conclusion regarding the suggestion to introduce nuclear power.

3.1.4. Challenges faced and lessons learned

The development of a comprehensive national position on the use of nuclear power including a policy and strategy for safety, is a significant step for any national government. The development of the national position and policy for safety covers a wide range of issues and involves extensive consultation with interested parties to obtain inputs and to build support for the position, along with an understanding of the special characteristics of nuclear power and the relevant international agreements and IAEA standards. Understandably, this exercise can take significant effort and time to reach maturity. In the case study countries, the factors that enabled the development of the national position and policy on safety included the active leadership of the government, effective engagement with interested parties, and the assistance of the IAEA through expert or peer review missions and in some cases technical assistance from other countries.

3.2. GOVERNMENTAL AND LEGAL FRAMEWORK FOR SAFETY

3.2.1. The issue

Principle 2 of SF-1 [1] states that "**An effective legal and governmental framework for safety, including an independent regulatory body, must be established and sustained.**" Many countries in the initial phase of consideration of a nuclear power programme do not have such a framework in place although certain elements may exist for other activities in the State such as the use of radiation sources in industry or medicine. Thus, in Phase 1 the government is to assess the need for new or amended legislation to support the safe and secure development of a nuclear power programme and make plans to enact the necessary legislation in Phase 2 if the decision is taken to proceed with a programme.

Paragraph 2.5 of GSR Part 1 (Rev. 1) [2] states

"[A]n effective governmental, legal and regulatory framework for safety … shall set out the following:

(1) The safety principles for protecting people — individually and collectively — society and the environment from radiation risks, both at present and in the future;
(2) The types of facilities and activities that are included within the scope of the framework for safety;
(3) The type of authorization that is required for the operation of facilities and for the conduct of activities, in accordance with a graded approach;
(4) The rationale for the authorization of new facilities and activities, as well as the applicable decision-making process;
(5) Provision for the involvement of interested parties and for their input to decision making;
(6) Provision for assigning legal responsibility for safety to the persons or organizations responsible for the facilities and activities, and for ensuring the continuity of responsibility where activities are carried out by several persons or organizations successively;
(7) The establishment of a regulatory body, as addressed in Requirements 3 and 4;
(8) Provision for the review and assessment of facilities and activities, in accordance with a graded approach;
(9) The authority and responsibility of the regulatory body for promulgating (or preparing for the enactment of) regulations and preparing guidance for their implementation;
(10) Provision for the inspection of facilities and activities, and for the enforcement of regulations, in accordance with a graded approach;
(11) Provision for appeals against decisions of the regulatory body;
(12) Provision for preparedness for, and response to, a nuclear or radiological emergency;
(13) Provision for an interface with nuclear security;
(14) Provision for an interface with the system of accounting for, and control of, nuclear material.
(15) Provision for acquiring and maintaining the necessary competence nationally for ensuring safety;
(16) Responsibilities and obligations in respect of financial provision for the management of radioactive waste and of spent fuel, and for decommissioning of facilities and termination of activities;
(17) The criteria for release from regulatory control;
(18) The specification of offences and the corresponding penalties;
(19) Provision for controls on the import and export of nuclear material and radioactive material, as well as for their tracking within, and to the extent possible outside, national boundaries, such as tracking of the authorized export of radioactive sources."

3.2.2. Points of interest

How the governments in the case study countries carried out the actions in Phase 1 recommended by SSG-16 (Rev. 1) regarding the legal, governmental and regulatory framework for safety, and what challenges they experienced and lessons they learned in doing so, namely:

"Action 20. The government should identify all necessary elements of a legal framework for the safety infrastructure and should plan how to structure and develop this framework.

"Action 21. The government should consider the process that should be employed to license nuclear facilities in the later stages of the programme."

In particular, in each case study country, which laws were enacted to ensure the establishment of an effective legal framework for safety? Did the legal framework clearly set out the functions and responsibilities of the organizations involved; particularly, those of the regulatory body? How did the legal framework complement or combine with the national policy?

3.2.3. Discussion

In all case study countries, a legal, governmental, and regulatory infrastructure for the safety of nuclear and radiation facilities and activities was in place before a nuclear power programme started. Egypt, Ghana, Nigeria, and Uzbekistan all have operated research reactors for many years along with the associated centres of expertise, in addition to regulating the use of radiation sources. Belarus had a long-standing legal arrangement in place for the uses of radiation sources in the country, as well as for the radiation safety of the public following the transborder impacts of the Chernobyl accident in a neighbouring State. There were research nuclear facilities as well. The UAE had extensive use of radiation sources in industry and medicine before initiating its nuclear power programme. Several countries reported that the experience and expertise gained through the regulatory oversight of these facilities and activities helped in developing the legal and governmental framework for their nuclear programme.

All countries reported that they aimed to prepare a comprehensive nuclear legal framework aligned with IAEA standards. In some cases, the assistance of the IAEA Office of Legal Affairs was enlisted directly in reviewing the draft nuclear laws.

Typically, the legal framework takes the form of a hierarchy, governed by the nuclear law and supported by other primary and secondary legislation. The specific form and organization of the legal framework in each country were influenced by prior experience and by the national legal traditions. In most countries, additional legislation related to the nuclear programme needed to be enacted or amended in addition to the comprehensive nuclear law.

Belarus, for example, employed a variety of legal instruments including, at the highest level, national laws, presidential decrees followed by government resolutions, and technical regulations. Belarus reportedly also adopted several technical regulations of the NPP vendor country where there were no equivalent national regulations, provided that the vendor regulations were in line with IAEA standards. Other importing countries also reported adopting this approach, including the UAE and Uzbekistan [18].

The UAE has also benefited from strong exchanges and support from the vendor country for the development of its regulatory requirements, safety assessment and licensing, inspection, and operating experience.

In Egypt, Law No. 7 of 2010 (as amended by Law No. 211 of 2017) [19] sets out the main elements of the nuclear framework for radiation protection, safety of nuclear facilities, emergency preparedness and response, radioactive waste and spent fuel management, transport of radioactive material, nuclear security, safeguards, import and export control, and civil liability for nuclear damage. Related legislation in Egypt includes:

- Law No. 209 of 2017 Establishing Executive Authority for supervising the construction of nuclear power plant projects for generating electricity [14];
- Law No. 210 of 2017 Amending Law No. 13 for Year 1976 for the establishment of the Nuclear Power Plants Authority [20];

- The Decree of the President of the Arab Republic of Egypt No. 196 of 1977 for Establishment of Nuclear Material Authority [14];
- Law No. 13 of 1976 for the establishment of the Nuclear Power Plants Authority [19].

Before construction of the El Dabaa NPP, Egypt's Nuclear Power Plants Authority and the vendor undertook a comprehensive review of the legal infrastructure that might impact the project. Several issues were identified, for instance tax and customs regulations, as well as labour law, that needed changes [14].

Similarly, in the UAE, Federal Law by Decree No. (6) of 2009 sets out the main elements of the legal framework for nuclear and radiation activities in the State [21] The mentioned law is complemented by Federal Law by Decree No. (4) of 2012 [22] concerning civil liability for nuclear damage, along with several Cabinet Resolutions describing fees for certain services and administrative penalties. Other related UAE laws include those that establish and govern key entities such as the Emirates Nuclear Energy Company (ENEC), or the national authorities for emergency preparedness and physical protection [23]. Also, the UAE has joined all the relevant international nuclear instruments in the fields of safety, nuclear security, safeguards, and nuclear liability.

Several countries developed their legal and regulatory framework over a period before starting the NPP project. For example, the Nigerian Nuclear Regulatory Authority (NNRA) was established by the Nuclear Safety and Radiation Protection Act in 1995 to regulate nuclear safety and radiological protection [24]. The latter act was revised in draft form following feedback from a review conducted by the IAEA.

The embarking countries assumed to enact the nuclear law by early phase 2, while the preparation and drafting started earlier. Some countries had enacted or drafted a comprehensive nuclear law and started the establishment of a national regulatory body to address safety, security and safeguards in parallel to conducting preliminary feasibility studies to support making the decision to initiate a nuclear power programme, which was reported as good practices [7]. Some countries experienced delays in development of the legal and governmental framework in a timely manner in line with the development of the nuclear power programme. In general, to maintain a high level of nuclear and radiation safety, the country's legal framework needs to undergo continuous improvements.

The recent Integrated Nuclear Infrastructure Review (INIR) mission concluded that Belarus has established a comprehensive legal framework through a series of laws, associated decrees, and regulations. Belarus is a party to most of the international legal instruments and has established the key organizations. In continuation of the work done Belarus needs to consolidate and strengthen its nuclear legislation and adhere to and implement the remaining international legal instruments to which Belarus is not yet a party.

Nigeria indicates that its updated nuclear law was passed by the national legislators and now awaits the approval of the president. Egypt and Uzbekistan also have national comprehensive nuclear law in place, and they are yet to become full parties to all the relevant international legal agreements, particularly the Convention on Nuclear Safety [14].

3.2.4. Challenges faced and lessons learned

All case study countries noted that the development of a comprehensive legal, governmental and regulatory framework is a complex process. Several countries indicated that they experienced challenges in completing the required framework on a timeline that was consistent with the goals they had set in line with the NPP project development, due to the time taken at various stages of assessing, drafting, approving, enacting and implementing laws. Another common theme across the case studies of the countries is the need to ensure consistency of the nuclear framework with international and national law.

The main lessons identified by the case study countries are as follows:

- Build on experience gained from the operation and regulation of other nuclear and radiation facilities in the country to identify the necessary elements of the legal infrastructure for safety.
- Study the experience of leading nuclear power countries in developing nuclear laws, together with consideration of the relevant IAEA safety standards.
- Assess the existing legislative framework of the country before the development of the nuclear law to ensure consistency and address conflicts.
- Follow up with the national parliament or legislative body to develop understanding of the specific nature of nuclear law and to provide any needed additional information to expedite the passage of the law.
- Use the NPP vendor country technical regulations, where practicable, to reduce the administrative burden of writing national regulations and to capitalize on the know-how of the vendor country.

3.3. REGULATORY INFRASTRUCTURE FOR SAFETY

3.3.1. The issue

The establishment of an independent regulatory body within the legal and governmental framework for safety is a fundamental principle set out in SF-1 [1] and in Articles 7 and 8 of the Convention for Nuclear Safety [11]. Therefore, countries that are considering embarking on a new nuclear power programme need to plan for the development of an appropriate regulatory infrastructure for safety. The regulatory infrastructure that exists to oversee any existing radiation and nuclear activities in the country can provide a basis of knowledge and experience to build on for the nuclear power programme.

3.3.2. Points of interest

How the governments in the case study countries addressed the actions in Phase 1 recommended by SSG-16 (Rev. 1) [3] regarding the regulatory infrastructure for safety, namely:

"Action 24. The government should recognize the need for an effectively independent and competent regulatory body and should consider the appropriate position of the regulatory body in the State's governmental and legal framework for safety.

"Action 25. The government should seek advice from the regulatory body on radiation safety issues relating to a nuclear power programme.

"Action 26. The government should identify the prospective senior managers of the regulatory body."

The following questions are also of interest:

- What was planned regarding the establishment of the regulatory infrastructure — both before and after a formal Government policy decision?
- How did the case studies countries build upon experiences acquired from regulating of nuclear facilities other than NPPs (e.g. research reactors, nuclear fuel cycle facilities) and non-nuclear activities (e.g. the use, handling and storage of radiation sources) to transition to NPP regulations with involvement of the regulatory body.
- How did the regulatory bodies in the case study countries establish and develop their functions and organizations? Were all the necessary elements of the regulatory framework in place in time to transition to Phase 2?
- The experiences of the case study countries in implementing these actions, and the challenges and solutions they encountered, are discussed below.

3.3.3. Discussion

In all six case study countries, the government recognized at an early stage the need for an independent and competent regulatory body to oversee the nuclear power programme and took corresponding action, as summarized below.

- Belarus created the Department for Nuclear and Radiation Safety (Gosatomnadzor) as a separate legal entity within the Ministry for Emergency Situations in accordance with the 2007 Presidential Decree No. 565 'On Several Measures Aimed at the Nuclear Power Plant Construction' [25]in Phase 2 several years before entering into an agreement with the Russian Federation for supply of the NPP.

- Egypt established the Egyptian Nuclear and Radiological Regulatory Authority (ENRRA) in 2012 following the passage of Law No. 7 of 2010 [19], supplemented by executive regulations issued by prime ministerial decree setting out the functions and responsibilities of the new regulatory body. The foregoing actions took place in Phase 2 several years before the first intergovernmental agreement in 2015 for supply of the El Dabaa NPP.

- Ghana established the Nuclear Regulatory Authority (NRA) under the Ministry of Environment Science Technology and Innovation (MESTI) through the passage of the Nuclear Regulatory Authority Act of 2015 [15].

- Nigeria established the Nigerian Nuclear Regulatory Authority (NNRA) by the Nuclear Safety and Radiation Protection Act of 1995 [24]. NNRA is developing safety infrastructure to support government policy in anticipation of future development of a nuclear power programme.

- The UAE nuclear programme moved on a relatively rapid timeline. Nevertheless, the UAE nuclear policy [16] issued in 2008 noted that "a primary objective, in the event the UAE chose to commission NPPs within its territory would be to establish a body authorized and competent to exercise supervision over nuclear safety independently of manufacturers and operators." Also, the nuclear policy [16] emphasized that "the establishment of an independent, vigilant and effective regulatory authority is a cornerstone for any stable, credible, safe and secure nuclear energy program". The UAE followed through on its policy objective with the enactment of Federal Law by Decree No. 6 of 2009 which sets out the legal framework for the conduct of peaceful nuclear activities in the State and establishes the Federal Authority for Nuclear Regulation (FANR) as the new independent regulatory body [23].

Uzbekistan, through the Atomic Energy Act of 2019 has established State Committee on Industrial Safety (Goskomprombez) as the official and authorized State body responsible for the implementation of a unified State policy and supervision in the field of radiation and nuclear safety at nuclear facilities [17]. This action was shortly after the signing of an intergovernmental agreement with the Russian Federation for supply of the NPPs. Most countries built on the experience and the regulatory infrastructure that was already in place for existing activities such as the use of radiation sources and nuclear research facilities in industry, science, and medicine, for instance as outlined in the following points.

- In Egypt ENRRA took over the responsibilities of the earlier National Centre for Nuclear Safety and Radiation Control (NCNSRC) that was created in 1991 for licensing and oversight of the ETRR-2 research reactor. In turn NCNSRC replaced the original Nuclear Regulatory and Safety Committee of the Egyptian Atomic Energy Authority (EAEA).

- Belarus also built its regulatory infrastructure for nuclear safety on elements that previously existed in the country. The creation of Gosatomnadzor followed the industrial, nuclear and radiation safety inspectorate (Gospromatomnadzor) that was formed many years earlier at the beginning of 1990s to carry out state supervision over industrial nuclear and radiation activities. At present, the Ministry for Emergency Situations in Belarus is responsible for regulation of nuclear and radiation safety via Gosatomnadzor and also for industrial safety through a separate department named Gospromnadzor. Uzbekistan has adopted a similar organizational structure in establishing the nuclear regulatory body.

- In Ghana, the NRA succeeded the Radiation Protection Bureau that had been created years earlier to supervise the operation of the GHARR-1 research reactor. The NRA was initially staffed by experienced staff and leaders who were seconded from the Ghana Atomic Energy Commission.

- Nigeria established the NNRA in 1995 to regulate radiation and nuclear safety including the NIRR-1 research reactor. Since then, a draft revision of the enabling legislation, the Nuclear Safety and Radiation Protection Act, has been prepared. This new version aims at strengthening the powers of the NNRA as the national nuclear regulator.

- The UAE, uniquely among the group of case study countries, did not have experience of operating a nuclear research reactor prior to launching its nuclear power programme and thus had relatively little existing safety infrastructure in place when it took the decision to embark on its nuclear power programme. A small number of experienced staff from another ministry involved in supervision of the uses of radiation sources were transferred to work in similar roles in the new regulatory body, FANR. In addition, FANR was able to recruit senior and highly experienced expatriate staff to support the activities of the newly established regulatory body.

Although none of the case study reports explicitly states when the senior managers of the regulatory body were identified or hired, the context shows that senior staff having prior experience of nuclear and radiation safety regulation were available in most of the countries to take up leadership positions in the new or reformed nuclear regulatory body. In the UAE's case, the absence of substantial nuclear skills in the country was addressed by the NEPIO hiring several senior managers from international sources before the regulatory body was officially established by law. In the case of Belarus, the indicator of importance of hiring the person at the position of the head of Gosatomnadzor is the fact that this appointment has to be agreed by the President of the Republic of Belarus.

3.3.4. Challenges faced and lessons learned

Several case study countries affirmed that their experience gained through regulation of research reactors and other nuclear facilities helped to identify and to establish the necessary elements of the regulatory framework for the NPP.

A major challenge reported by several countries lies in managing the rapid growth of the regulatory body and developing the competence of its staff while simultaneously performing the required functions such as developing regulations, performing safety assessments, licensing, and inspection in line with the NPP implementation schedule. The use of external experts, or technical support organizations (TSOs), and establishing cooperation agreements with experienced regulatory bodies in the vendor country or third countries, are common strategies used to manage change and to develop regulatory competence.

A further challenge reported is the need for the regulatory body to prioritize and develop technical regulations. This challenge was magnified in some cases by the absence of a decision on the NPP technology. On one hand, in cases when the NPP technology is known, some countries, notably Belarus and the UAE, report on having adopted some of the regulations or standards of the vendor country regulator where these are compatible with the national framework. In other cases where the NPP technology has not yet been decided, technology neutral regulations have been developed based on the corresponding IAEA standards.

3.4. OTHER GOVERNMENTAL CONSIDERATIONS – GLOBAL NUCLEAR SAFETY REGIME

3.4.1. The issue

The term 'global nuclear safety regime' refers to the institutional, legal and technical measures for ensuring the safety of nuclear installations throughout the world. As stated in para. 3.2 of GSR Part 1 (Rev. 1) [2]:

"The features of the global nuclear safety regime include:

(a) International conventions that establish common obligations and mechanisms for ensuring protection and safety;
(b) Codes of conduct that promote the adoption of good practices in the relevant facilities and activities;
(c) Internationally agreed IAEA safety standards that promote the development and application of internationally harmonized safety requirements, guides and practices;
(d) International peer reviews of the regulatory control and safety of facilities and activities, and mutual learning by participating States;
(e) Regular multilateral and bilateral cooperation between the relevant national and international organizations to enhance safety by means of harmonized approaches as well as to increase the quality and effectiveness of safety reviews and inspections, by means of sharing of knowledge and feedback of experience."

National participation in the global nuclear safety regime has implications for the legal, governmental and regulatory framework due to the obligations that countries agree to by becoming parties to certain international legal instruments.

3.4.2. Points of interest

Of interest are how the governments of the case study countries addressed the actions given by SSG-16 (Rev. 1) [3] for Phase 1 regarding the Global Nuclear Safety Regime [26], namely:

> **"Action 11. The government should prepare for participation in the global nuclear safety regime.**

> **"Action 12. The government should begin a dialogue with neighbouring States with regard to its projects for establishing a nuclear power programme.**

> **"Action 13. The government and relevant organizations (if such organizations exist) should establish contact with organizations in other States and international organizations to seek advice on safety related matters."**

Also of interest are the following questions:

- How did the regulatory body contribute to participation in the global nuclear safety regime?
- How did the regulatory body establish bilateral agreements with other countries?
- What plans are in place to invite IAEA missions (e.g. INIR, IRRS)?

The responses of the case study countries, together with the challenges encountered and lessons learned are assessed in the following subsection.

3.4.3. Discussion

All the case study countries reported that the national government during its planning for the nuclear programme developed an awareness of elements of the global nuclear safety regime and the corresponding obligations. The level of insight and preparedness was enhanced by experience gained from existing nuclear and radiation activities in the country including the operation of research reactors. Actions taken to develop national participation in the global nuclear safety regime included becoming a party to the relevant international instruments, the use of IAEA standards as references in developing regulations, participating in technical cooperation programmes with the IAEA, hosting peer review missions as appropriate to each Phase of the nuclear programme, and engaging in regional and international cooperative fora. The table below shows the international instruments to which each case study country is currently a party.

In some cases, the government is acting to become a party to additional international agreements. For instance, Egypt has set up a national committee for studying the feasibility of joining relevant international legal instruments and is giving priorities to the ratifying of the Convention on Nuclear Safety and the Convention on Supplementary Compensation for

Nuclear Damage[2] Uzbekistan reports that it is working toward joining relevant instruments including the Vienna Convention on Civil Liability for Nuclear Damage, the Convention on Early Notification of a Nuclear Accident, the Convention on Nuclear Safety, and the Convention on Assistance in the Case of a Nuclear Accident or Radiological Emergency.

TABLE 1: CASE STUDY COUNTRIES PARTICIPATION IN SAFETY-RELATED INTERNATIONAL INSTRUMENTS[3]

International Instrument	Belarus	Egypt	Ghana	Nigeria	UAE	Uzbekistan
Convention on Early Notification of a Nuclear Accident [27]	X	X	X	X	X	
Convention on Assistance in the Case of a Nuclear Accident or Radiological Emergency [28]	X	X	X	X	X	
Convention on Nuclear Safety [11]	X		X	X	X	
Joint Convention on the Safety of Spent Fuel Management and the Safety of Radioactive Waste Management [12]	X		X	X	X	X
Convention on the Physical Protection of Nuclear Material [29]	X		X	X	X	X
Amendment to the Convention on the Physical Protection of Nuclear Material [30]	X		X	X	X	X
Vienna Convention on Civil Liability for Nuclear Damage [31]	X		X	X		
Protocol to Amend the Vienna Convention on Civil Liability for Nuclear Damage [32]	X		X		X	
Convention on Supplementary Compensation for Nuclear Damage [33]					X	
Joint Protocol Relating to the Application of the Vienna Convention and the Paris Convention [34]		X	X		X	
Comprehensive Safeguards Agreement (CSA) — based on the Treaty on the Non-Proliferation of Nuclear Weapons (NPT), and the Structure and Content of Agreements Between States Required in Connection with	X	X	X	X	X	X

 More information on the legal framework of the country is available at https://www.iaea.org/resources/legal/country-factsheets

[3] The information presented in the table was updated to reflect the status of the international legal instruments in the case study countries as per the time of publication of the TECDOC. However, the accompanying text reflects the situation as of July 2023, which was agreed upon as the reference point for the TECDOC reporting project progress in accordance with Section 2

the Treaty on Non-Proliferation of Nuclear Weapons [35]						
Additional Protocol to CSA [36]			X	X	X	X
Revised Supplementary Agreement Concerning the Provision of Technical Assistance by the IAEA [37]	X	X	X	X	X	X

A few countries in this study reported having consulted neighbouring States that might be affected by a nuclear installation in accordance with Articles 16 and 17 of the Convention on Nuclear Safety. Ghana indicated that after a comprehensive report on the proposed nuclear power programme was submitted to the Government, an interministerial committee was formed by the President comprising the Minister of Foreign Affairs and Regional Integration, Minister of Finance, and Minister of Environment Science Technology and Innovation to inform neighbouring countries of Ghana's intention to construct an NPP.

Uzbekistan has also reported the conclusion of an agreement with the Russian Federation for the prompt notification of a nuclear accident and exchange of information in the field of nuclear and radiation safety. Additionally, Belarus has indicated the establishment of bilateral agreements on assistance and exchange of information and notifications with neighbouring countries including Latvia, Lithuania, Poland, the Russian Federation and Ukraine [38].

All case study countries report that they have established cooperation agreements with other countries and international organizations and are participating in regional and international networks for exchange of information and assistance.

3.4.4. Challenges faced and lessons learned

Several case study countries mentioned that their participation in the various elements of the global safety regime in the initial and subsequent phases of their programme strongly supported the development of the nuclear and radiation safety infrastructure.

Belarus emphasized that interaction with the organizations of the NPP supplier country at different levels is also of vital importance. On the other hand, Nigeria noted that the absence of a decision on the NPP technology makes cooperation with countries with experience in nuclear licensing more challenging.

Participation in the global safety regime by each country requires adequate resources and knowledge in the regulatory authority, the operating organization, government ministries for foreign affairs and other involved organizations. It is important that such resources are made available for this purpose. Another factor highlighted by Nigeria is the need for engagement of the legislature and senior government officials to highlight the necessity and the benefits of ratifying international treaties and participating in international cooperation activities. The regulatory body may possess relevant knowledge but typically is not empowered to make commitments on behalf of the national government, this being the function of the executive and the legislative branches.

4. KEY PRIORITIES FOR THE GOVERNMENT AND THE REGULATORY BODY IN PHASE 1

4.1. SITE SURVEY

4.1.1. The issue

The selection of a site for a nuclear installation has important implications for safety. The natural and human induced external hazards associated with the site must be considered for the site suitability, site parameters for the design of the plant and environmental impact assessment, and some site characteristics may affect the planning of effective emergency response actions [39]. To demonstrate protection of people and the environment from harmful effects of ionizing radiation, the site survey process is typically carried out in stages: regional analysis, starting with investigation of a large region to identify candidate sites; assessment and ranking of the potential sites; and screening test, by screening unsuitable potential sites to identify a number of candidate sites.

Paragraph 2.3 of IAEA Safety Standards Series No. SSG-35, Site Survey and Site Selection for Nuclear Installations [40] states that 2.3. "The siting process and the site evaluation process include five different stages. The siting process for a nuclear installation consists of the first two stages of these five. In the site survey stage, large regions are investigated to find potential sites and to identify one or more candidate sites. The second stage of the siting process is site selection, in which unsuitable sites are rejected, and the remaining candidate sites are assessed by screening and comparing them on the basis of safety and other considerations to arrive at the preferred candidate sites."

Site selection, site characterization and site evaluation might need more detailed survey, data gathering and analysis results for the justification of the selected site in later phase.

4.1.2. Points of interest

Of interest are how the governments of the case study countries addressed the action below given by SSG-16 (Rev. 1) [3] for Phase 1 regarding siting, their involvement in the site survey, and what safety requirements each regulatory body put in place early in the process:

> **"Action 160. The government should ensure that potential sites are identified, and candidate sites are selected on the basis of a set of defined criteria, at a regional scale and with the use of available data."**

Further, if the regulatory body is established in Phase 1, what was the level of its involvement in the site survey? And what arrangements has each regulatory body put in place for establishing safety requirements for site evaluations (early in the process prior to site selection)?

4.1.3. Discussion

Most case study countries completed surveys of potential candidate sites by the time of, or soon after, the decision to proceed with the nuclear power programme that marks the transition from Milestones Phase 1 to Phase 2 [4]. In Egypt, the NPPA completed studies leading to the identification of candidate sites many years before the present El Dabaa project began. Belarus identified the candidate sites by the time the decision was taken to proceed with the programme, as did Uzbekistan.

The UAE's identification of candidate sites was complete within a year of the programme launch. Ghana and Nigeria are currently conducting siting studies in preparation for the planned launch of their NPP project. Table 2 shows the timing of key actions regarding siting in all the case study countries.

TABLE 2: TIMING OF KEY ACTIONS REGARDING SITING IN THE CASE STUDY COUNTRIES

Country	Decision on nuclear power programme[4]	Siting requirements issued	Regional survey and selection of candidate sites	Site licence
Belarus[41]	2008	Law on use of Atomic Energy 2008, Technical codes TCP 97/2007 – 99/2007 and 101/2007-102/2007	2008	2012
Egypt [14]	2007	Site Evaluation Requirements for Nuclear Installations 2009/ updated 2016	1978	2019
Ghana [42]	TBD	Under development by NRA	2017	TBD
Nigeria [43]	TBD	Regulation on Licensing of Site for NPP in 2021	2007-2019	TBD
UAE [23]	2008	FANR draft regulations 2010	2009-2010	Final approval with construction licence 2012
Uzbekistan [44]	2018	Resolution of the Cabinet of Ministers No. 390 of June 17, 2020	2018	TBD

In several countries the site surveys to identify candidate sites were led by the future operating organization, for instance NPPA in Egypt, ENEC in the UAE, and Agency for the Development of Nuclear Energy (Uzatom) in Uzbekistan. In other countries, State institutions performed this task, as in Belarus, where a State Commission established for the purpose led work by the National Academies of Sciences. In Ghana, the Ghana Nuclear Power Programme Office provided a comprehensive siting charter while the geological and seismological assessment was carried out with the involvement of bodies having relevant expertise including the Ghana Geological Survey Department, the Minerals Commission, the Earth Science Department of the University of Ghana, and the Water Research Institute.

The requirements for site evaluation in some cases were issued before the survey of candidate sites. For instance, in Belarus, the Law on the Use of Atomic Energy and technical standards for siting were available around the time of the site survey. In other cases, the requirements were formally issued after the candidate sites were identified. ENRRA in Egypt issued siting requirements in 2009 and updated them in 2016. FANR also issued some relevant regulations toward the end of the period of the site survey. Nigeria has reportedly issued Safety Regulations

---

[4] The decision on a nuclear power programme by any country as marked by the conclusion of an intergovernmental agreement and/or a supply contract.

for Licensing of Site for Nuclear Power Plants, while in Ghana the regulations are still under development. According to the Milestones approach [4], early in Phase 2, the regulatory body will need to define siting requirements to be taken into account by the owner/operator in the final site selection and characterization(see SSG-16 (Rev. 1) [3]), to ensure that the regulatory requirements were considered during the site surveys. The case study reports give insufficient detail to enable analysis of this point. However, if the organizations that conducted the site survey followed the relevant IAEA safety standards as well as those of the vendor country, it is likely that the work would be readily reconciled with the regulatory requirements.

Although SSG-16 (Rev. 1) [3] indicates that the site licence is granted in principle during Phase 2 of the programme, in fact the final regulatory approval of the NPP site was granted at different points in the case study countries. In Belarus, the regulatory body Gosatomnadzor issued the site licence before the contracts were issued for the NPP, thus formally in Phase 2. In Egypt, ENRAA issued the site licence in Phase 3 after the NPP contracts were placed. In the UAE, the operator ENEC formally submitted the site characterization results as part of the safety analysis report (SAR) for the construction licence application in Phase 3 and FANR subsequently granted authorization through the construction licence.

Several countries reported using IAEA services related to siting, including Site and External Events Design (SEED) review missions and various training courses and workshops to support the development of staff competence and the conduct of the site safety review and evaluation exercise.

4.1.4. Challenges faced and lessons learned

A challenge common to several countries is the acquisition of the needed competencies to perform and to assess the adequacy of the NPP site. In some countries such as Belarus and Ghana, institutions existed within the country with the capability to do this work. Other embarking countries had to develop expertise. It is a particular challenge for the newly formed regulatory bodies to ensure that the timescale of recruiting staff, developing competence, issuing regulations, and reviewing the site characterization is in line with the NPP project.

Responses to this challenge included engaging IAEA services for training and expert missions, hiring TSOs, using existing expertise in other organizations in the country, cooperating with other experienced regulatory bodies, and deferring the final site approval until a later phase in the programme, typically early in Phase 3.

4.2. CONSULTATION WITH INTERESTED PARTIES

4.2.1. The issue

There is a growing recognition of the need for transparency and openness in nuclear and radiation safety matters, aiming to build trust and communication between the regulatory body and interested parties. Requirements by the regulatory body to communicate with interested parties are embedded in the Fundamental Safety Principles set out in SF-1 [1] and in several IAEA safety standards.

4.2.2. Points of interest

Of interest are how the governments of the case study countries addressed the actions shown below in SSG-16 (Rev. 1) [3] for Phase 1, what mechanisms the regulatory bodies set up to involve the public and other interested parties, which interested parties the regulatory bodies engage with, and the issues on which they are consulted and informed:

"**Action 39: The government should establish a policy and guidance to inform the public and other interested parties of the benefits and risks of nuclear power to facilitate their involvement in the decision-making process on a prospective nuclear power programme.**

"**Action 40: The government should establish a process to ensure that the comments arising from consultation with the public and other interested parties are considered, and it should communicate the results of these considerations to the interested parties.**"

4.2.3.   Discussion

In every case study country, the government clearly recognizes the need for transparency and openness in its approach to planning and development of the nuclear programme. In each country the high-level requirement for communication and consultation with interested parties is set out in law: either in the nuclear law, or in a related instrument such as a government resolution. Recommendations on this topic are provided in IAEA Safety Standards Series No. GSG-6, Communication and Consultation with Interested Parties by the Regulatory Body [45].

In Belarus, several laws, regulations, and other instruments direct interested parties' involvement activities. The 2008 Law on the Use on Atomic Energy sets out the main responsibility for engaging the public and other interested parties for nuclear power. It provides that the government determines the cases and procedure of discussion of questions in the field of use of atomic energy with participation of non-governmental organizations, other organizations and citizens [41].

Belarus reportedly set up formal coordination mechanisms among the key State institutions. Interaction between government authorities and State organizations in the early stages of the nuclear power programme took place within the framework of established practice in the State. With further implementation in Phase 2, two coordination mechanisms were created by the resolutions of the Government, namely an Inter-ministerial Commission headed by the Vice Prime Minister, and a working group for coordination of State supervision of construction headed by the Deputy Minister for Emergency Situations.

The Ghana Nuclear Power Programme Organization is also reported to play a leading role in communication and consultation with interested parties regarding the planned nuclear programme, supported by the efforts in this direction of the NRA. The Nuclear Regulatory Authority Act 895 of 2015 requires the NRA to educate the public on nuclear and radiation matters, to engage with interested parties, and to provide regular and emergency communications.

Other cases in this study concentrate on the role of the regulatory body in interested parties' engagement, as summarized in point form below. However, it would be kept in mind that the government and the NPP owner/operator, according to. IAEA Nuclear Energy Series No. NG-G-5.1, Stakeholder Engagement in Nuclear Programmes, para. 4.2 states that" The owner/operator develops public information activities and materials to facilitate the public's understanding of the project" [46]. They also have their respective roles to play in communication and consultation with other interested parties.

- In Egypt, the Law on Regulation of Nuclear and Radiological Activities, its amendments, and the Executive Regulations make ENRRA responsible for informing the public about the regulatory process of nuclear and radiation activities, establishing the means to involve the public, taking steps to disseminate nuclear safety and security culture, and providing the public with information about the status of nuclear and radiation safety in the area of their residence (unless this information is confidential).
- In Nigeria, the Nuclear Safety and Radiation Protection Act empowers the NNRA to provide training, information and guidance on nuclear safety and radiation protection.
- The UAE Nuclear Law requires FANR to maintain the highest standards of transparency while performing its functions and to facilitate the public's access to all relevant information on its activities.

Given their legal mandates, the regulatory bodies in most case study countries have established mechanisms for engagement of interested parties, as summarized below:

- Gosatomnadzor in Belarus has adopted and maintained an Information and Communication Strategy with the objective of developing and maintaining a constructive dialogue with public authorities, organizations, experts, media and the public (in Belarus and internationally) and to achieve a common understanding of the issues of nuclear and radiation safety, the role and functions of the regulatory body, and the development of a safety culture in the nuclear field.
- In Egypt, ENRRA has developed and maintained a communications strategy that its communications team has updated from time to time based on analysis of experience feedback. ENRRA uses various communications tools and techniques in different phases and situations to achieve its communications goals, the most important of which include in-person outreach activities, media relations, public hearings and meetings, and digital presence through different channels.
- The NRA in Ghana is developing a Strategic Communications Plan, an International Relations Policy and an Emergency Communications Plan. A Stakeholder Mapping and Engagement Committee has been formed with representation from the Directorates of the NRA to identify the relevant interested parties and to plan engagement activities. These plans have been scrutinized through various peer reviews.
- The NNRA in Nigeria has developed a Communication Policy and Strategy, which received management approval in 2021, that covers external and internal means of communication between the NNRA and other interested parties, as well as the means for monitoring and evaluation of the effects of such communication. While internal communication targets NNRA staff and members of the governing board and aims at strengthening the organizational safety culture, external communication promotes knowledge and awareness of the vision, mission, mandates, activities, and services of the NNRA to interested parties.
- To fulfil its transparency obligations under the Nuclear Law and align with the overall Federal Government Communication Strategy, FANR in the UAE has developed and implemented a communication strategy and programmes aimed at various interested parties including the public, Government entities, international organizations, licensees, students, media and employees. FANR employs different channels and tools to engage and communicate with its interested parties as the website, media releases, publications

and reports, meetings with interested parties, and consultation on draft regulatory documents.

The external interested parties with whom the national regulatory bodies reportedly engage may be understood as falling in two broad categories of 'statutory interested parties' and 'non-statutory interested parties. [43]

Statutory interested parties are those organizations and bodies required by law to be involved in any planning, development or operational activity plus those that will be directly or indirectly impacted. Such statutory interested parties thus include international organizations, national and local government bodies involved in policy making and implementation, licensees, and authorities in various areas such as land use and planning, environmental protection, electric distribution, and emergency planning that have a supporting role in the development of a nuclear facility.

Non-statutory interested parties include those organizations and individuals who feel themselves affected by in some way by the nuclear programme. Some interested parties in this category may be self-selected. The general public, local communities and non-governmental organizations (NGOs) fall into this group. Their adequate inclusion may contribute significantly to the success or failure of a nuclear project.

As noted above, the principal matters on which the national regulatory bodies engage their interested parties include:

- Provision of information about the mandate, the activities, and the decisions of the regulatory body.
- Development of a common understanding of the issues of nuclear and radiation safety.
- Provision of information about the status of nuclear facilities and activities.
- Consultation on regulatory initiatives such as drafting regulations or licensing a facility.
- Promotion of a safety culture.

4.2.4. Challenges faced and lessons learned

The case studies show that the governments in most of the countries involved clearly recognized the importance of engaging interested parties throughout the planning and implementation of the national nuclear power programme. In most countries studied, the regulatory body has a legal responsibility to engage and consult with interested parties. Given these legal mandates, the regulatory bodies have implemented interested party engagement strategies with multiple components aimed broadly at providing information on regulatory activities, consulting with interested parties, and seeking their input on regulatory decisions. A few countries also reported on the activities of the government in this area. It was kept in mind that both the government and the owner/operator of the NPP have respective roles and responsibilities for interested party engagement complementing those of the regulatory body.

4.3. LEADERSHIP AND MANAGEMENT FOR SAFETY

4.3.1. The issue

Principle 3 in SF-1 [1] states that "**Effective leadership and management for safety must be established and sustained in organizations concerned with, and facilities and activities that give rise to, radiation risks**". Leadership and management for safety, an integrated management system and a systemic approach in all relevant organizations are all essential to the achievement of adequate levels of safety measures and the fostering of a strong safety culture. Requirements on this topic are established in IAEA Safety Standards Series No. GSR Part 2, Leadership and Management for Safety [47].

4.3.2. Points of interest

Of interest are how the case study countries addressed the actions below given by SSG-16 (Rev. 1) [3] for Phase 1:

> **"Action 72: The government should take into account the essential role of leadership and management for safety to achieve a high level of safety and to foster a strong safety culture within organizations.**

> **"Action 73: The government should ensure that all the activities conducted are included within the framework of an effective integrated management system.**

> **"Action 74: The government, when identifying senior managers for the prospective organizations to be established, should look for people with leadership capabilities and an attitude that emphasizes safety culture."**

The following questions are also of interest. What common features and differences are observed among the arrangements each regulatory body put in place for its management system? How did the regulatory bodies of the case study countries manage the development of their management systems? When was the development of the management system completed in relation to Phase 1 and transitioning to Phase 2? How was safety culture fostered? How were the senior managers identified, selected and assigned to the regulatory body?

4.3.3. Discussion

In all the case study countries, the government has considered the importance of leadership and management for safety. During Phase 1 of the nuclear power programme the government's action took various forms in different countries, such as the creation of interdepartmental coordination mechanisms for implementation of the nuclear power programme in Belarus; the establishment of similar coordination arrangements in Egypt; the development of guidance for implementation of management systems by the Ghana Nuclear Power Programme Organization (GNPPO); and the adoption of a policy commitment to the highest standards of safety and security by the UAE.

In those case study countries that have progressed to Phases 2 and 3, the owner/operator and the regulatory body — which are the key organizations with responsibilities for safety in the nuclear programme — have developed management systems and a safety culture in their respective organizations.

In several countries, the regulatory body has issued requirements that apply to the management system and safety culture of the owner/operator. The owners/operators have developed their management systems to fulfil their prime responsibility for safety and to conform with the regulatory requirements, as follows:

- In Belarus, to ensure nuclear and radiation safety, Gosamtomnadzor developed and put into effect general requirements for the operating organization's management systems. These rules establish general technical requirements for the constituent elements, the procedure for the functioning and improvement of administrative systems of the operating organization in terms of ensuring nuclear and radiation safety at all stages of the lifetime of nuclear facilities. The Belarusian NPP has committed to a Safety Policy and has developed a management system in accordance with GSR Part 2 [48] as well as national standards which is integrating various elements including policies, processes, and quality assurance programmes.

- In Egypt, the ENRRA issued in 2016 the requirements for management system at regulated facilities and activities. The NPPA had begun the development of its management system in 2012 covering NPP project management and operation and maintenance, and will integrate programmes for safety, health, environment, quality, societal and economic elements, as specified in GSR Part 2 [14].

- In the UAE, FANR issued regulations and regulatory guides setting out requirements for the management system of the operating organization. Nawah Energy Company (Nawah), the operating subsidiary of ENEC, has responded by developing a comprehensive management system for operation of the Barakah NPP that implements Nawah's vision, mission and values through various elements including performance indicators, strategic 5P operating model, (Purpose, Principles, Process, People and Performance) programmes, processes, procedures, business plans and model of governance [49].

The regulatory bodies have developed and implemented management systems in their own organizations to ensure that their regulatory responsibilities are properly discharged, to maintain and improve their performance through planning, control, and supervision of their safety related activities, and to foster a safety culture through the development of leadership as well as good attitudes and behaviour in relation to safety on the part of individuals and teams. The regulatory activities are summarized below.

- Gosatomnadzor appointed a Coordination Council in 2020 for coordinating the development, implementation, and maintenance of the management system in response to an IAEA IRRS mission finding. The current integrated management system (IMS) manual identifies and describes processes in three categories, specifically management processes, core processes and support processes, along with implementing procedures and other IMS documents. Gosatomnadzor has conducted training on the use of the IMS for staff with the support of the IAEA, European Union (EU) and Rosatom and has performed the first safety culture self-assessment [50].

- In Egypt, ENRRA in relation to transitioning to Phase 2, has established a management system. The Management System Manual outlines ENRRA's vision, mission and policy, as well as a description of the structure of the management system, the management system processes, the organizational structure and functions, the importance of leadership and safety culture, allocation of resources and training, and departmental objectives. ENRRA management gives priority to the development and foster a safety culture at all levels of the organization and aims to create an atmosphere of understanding and commitment of all personnel to safety. ENRRA has appointed a responsible person, has prepared a procedure for creation, development, and maintenance of safety culture, and draws up annual plans for safety culture.

- The NRA in Ghana is documenting the processes and procedures in use at the NRA to form the base of an integrated management system in line with GSR Part 2 [16], with support from the IAEA and the EU. A team of seven representatives from all Directorates of the NRA is leading the development. The NRA is also seeking to comply with ISO 9001:2015 for Quality Management Systems.[51] The NRA's management system builds on procedures inherited from the former regulatory body and is also draws upon experiences from other regulatory bodies in countries such as Canada, Egypt, Morocco, Lithuania, the Kingdom of the Netherlands, Pakistan and Slovenia. Self-assessments of the management system and training have been conducted with EU assistance.

- The NNRA in Nigeria developed a management system manual. In 2017, an IAEA IRRS mission reviewed the NNRA management system and made recommendations for improvement. A team was convened to review the integrated management system manual, with representation from each department and some strategic divisions and units along with support from IAEA knowledge networks such as African Regional Co-operative Agreement for Research, Development and Training Related to Nuclear Science and Technology (AFRA), the Forum of Nuclear Regulatory Bodies in Africa (FNRBA), Regulatory Cooperation Forum (RCF), and technical cooperation (TC) projects providing regional training and workshops. The NNRA is also receiving support for assessment of its management system from the EU.

- Pursuant to the commitments set forth in the UAE Nuclear Policy, FANR has implemented an IMS according to GSR Part 2 [49] . FANR's IMS integrates requirements for safety, security and safeguards, sets out the organization's mission, vision and core values, and incorporates structured policies, processes and procedures that enable FANR to deliver its functions effectively and support the development of a strong safety culture. FANR's IMS Committee is chaired on a rotational basis by one of the two Deputy Directors General and is composed of all FANR directors, who meet regularly to openly discuss the implementation and operation of the management system.

- In Uzbekistan, Goskomprombez's management system includes charters of its four Directorates and their 14 departments. Internal procedures are under development in accordance with annually approved schedules. Goskomprombez is reportedly planning to develop and implement an IMS. Consultations were carried out with an external TSO regarding its development, a draft resolution was developed and submitted to the Cabinet of Ministers for approval [17].

4.3.4. Challenges faced and lessons learned

Some countries indicated that the management of a large and rapidly evolving NPP project is a challenge, considering the need to perform functions in one phase while simultaneously preparing for the next phase. Challenges arising for safety culture can include:

- Maintaining the focus on safety while dealing with other interested parties.

- Considering economic factors and budget limitations and keeping appropriate resources.

- Interfacing with other regulations and regulatory bodies.

Several regulatory bodies mentioned that although the IMS needs to be developed in the early stages of the organization of the regulatory body, such a task may be unfamiliar for a new regulatory body; the staff may lack specific knowledge and skills, and it may take a long time.

Some solutions that were identified to address these challenges include:

- Treating the nuclear programme as a major national project from the very beginning has a positive impact on leadership, management, and safety culture.

- Leaders at all levels in an organization demonstrate priority to safety.

- Open and transparent communication with all interested parties by the regulatory body builds trust and confidence concerning regulatory activities and decision making.

- Clear regulatory frameworks and regulatory focus on the licensee meeting its prime responsibility for safety.

- Self-assessment, learning and continuous improvement including arrangements for formal training, implementing best practices, and participating in international activities.

- Support from the IAEA, other experienced regulatory bodies, and external experts.

4.4. DEVELOPMENT OF HUMAN RESOURCES

4.4.1. The issue

Introducing nuclear power requires a broad range of knowledge and skills including management and administrative skills as well as technical skills across various scientific and engineering disciplines. Specific areas of expertise are needed including reactor physics, accident analysis and nuclear materials science. This applies to the regulatory body, the owner/operator, TSOs and other relevant organizations. To build up national capabilities in an embarking country, a strategy is needed for the development and retention of skilled human resources to meet the needs of in the different phases of the nuclear power programme.

4.4.2. Points of interest

Of interest are how the countries in this case study addressed the following actions given by SSG-16 (Rev. 1) [3] for Phase 1:

"Action 85: The government should consider a strategy for attracting, recruiting, training and retaining an adequate number of experts to meet the needs of all organizations involved in ensuring safety in a prospective nuclear power programme.

"Action 86: The government should identify competences required in areas relating to nuclear safety and the approximate number of experts needed.

"Action 87: The government should identify national institutions and institutions in other States that could provide education and training and could start training in key areas relating to nuclear safety.

"**Action 88: The government should identify gaps in safety related training at existing training institutions and should plan to strengthen existing training institutions or to establish new training institutions to fill these gaps.**

"**Action 89: The government should ensure that prospective senior regulators identified by the government and prospective safety experts to be involved in the nuclear power programme gain an understanding of the principles and criteria of nuclear safety.**"

Further, the following points are of interest for HRD: In each country, how did the new regulatory body plan for and implement the activities related to HRD? What sources were used for HRD (internal and/or external)? Were the human resources in place in time to perform the needed regulatory activities in Phase 2? What use was made of TSOs/external support organizations to support in-house capability? What IAEA resources/tools were used (e.g. TC, RCF)?

4.4.3. Discussions

The reports from case study countries provide a range of information about the actions taken by government in Phase 1. Some reports addressed only the regulatory body whereas the scope of action for HRD recommended by IAEA Safety Standards Series No. GSG-12, Organization, Management and Staffing of the Regulatory Body for Safety [52] and set out in IAEA Nuclear Energy Series No. NG-T-3.10, Human Resource Management for New Nuclear Power Programmes [53] covers the NEPIO, the regulatory body and the owner/operator. Additional sources, mainly reports of recent IAEA missions, have been consulted to supplement the information in the case study reports.

Several countries reported that the government implemented a national strategy for human resources for the new nuclear power programme. The national programmes addressed the overall needs of the key organizations, namely the owner/operator and the regulatory body, and provided for coordinated actions to develop the national education and training infrastructure to satisfy those needs. Examples include:

- Belarus reported that the Government established a national training system at the start of Phase 2 to support the nuclear power programme with highly qualified specialists and to maintain an adequate level of knowledge for safe, reliable and efficient NPP operation. The National Staff Training Programme for Nuclear Energy of the Republic of Belarus was approved by the Council of Ministers of the Republic of Belarus in 2008 covering the period 2008 to 2020. The programme aims to organize the training system for the acquisition of necessary knowledge and skills for NPP construction, safe operation, and nuclear and radiation safety. This training system includes a set of organizational and technical actions by the state administration bodies, higher and secondary specialized educational institutions, industrial and technical schools, and other public institutions.

- Egypt has long experience in the application of nuclear technologies, and this provided a basis for the identification of the workforce needed for the nuclear power programme. The licensing and inspections of research reactors has resulted in several personnel having experience and expertise in safety of nuclear installations. This experience helped to identify the necessary competences and number of staff and to assess gaps in safety related training. An Education and Training Committee was formed to coordinate the integration of the HRD plans among the key organizations. The Committee is responsible for: (a) formulating strategies for training, education, and specialized national needs; (b) offering opinions on agreements and other forms of cooperation with foreign bodies; and (c) maximizing the benefits from the available infrastructure at national nuclear authorities and universities [14].

- In Ghana, a national human resource plan for the nuclear power programme has been developed with input from all key organizations and was approved by the GNPPO Board in June 2019, i.e., in Phase 2 [54].

- From the beginning of the nuclear power programme, the Government of the UAE recognized the importance of human resources. The UAE Nuclear Policy published in Phase 1 outlines the basis for a strategy to develop the needed human resources to regulate, manage, operate and maintain the safety of nuclear facilities. The national human resources strategy included two tracks, namely, staffing by experienced international staff to address the immediate and mid-term needs, and development of national capacity to ensure long-term sustainability. This strategy resulted in the recruitment of personnel having previous experience and skills in NPP regulation, construction and operation in the early phases of the programme.

  A national capacity building effort was implemented by ENEC and FANR in collaboration with Khalifa University and other educational and vocational training institutions in Phase 2. These entities worked together to identify the overall human resource needs and to implement education, training, and recruitment programmes to satisfy these needs. Each organization, in addition, developed its own specific human resources strategy.

  Khalifa University in partnership with Sandia National Laboratories and Texas A&M University established the Gulf Nuclear Energy Infrastructure Institute (GNEII) which provided an institutional capability for nuclear energy infrastructure development emphasizing education in nuclear energy safety, safeguards and security and which many UAE staff attended.

- In Uzbekistan, a strategy for HRD for the nuclear power programme was adopted by Presidential Resolution in October 2019, i.e. in Phase 2. All key organizations participated in developing the national HRD plan. According to the Presidential Resolution, during the period 2019–2030 continuous training of personnel will be provided for all organizations involved in the implementation of the nuclear power programme and measures will be taken to retain industry specialists. The Minister of Energy is responsible for coordinating of the strategy, and the heads of relevant State bodies are responsible for its implementation.

  The strategy emphasizes the importance of safety culture and of having highly qualified personnel at the NPP, Goskomprombez and other State bodies involved in the nuclear power programme. It addresses the short- and long-term staffing needs of all organizations supporting the nuclear power programme including TSOs like the Academy of Sciences.

Regarding the human resources (HR) needs of the operating organization specifically, all countries that have actively implemented a nuclear power programme have identified the organization and staffing needs for the operator and have established arrangements for recruitment, education and training of people to meet these needs. In most cases, the necessary education and training is provided partly on the national education and training infrastructure, partly by the owner/operator, and partly on training supplied by the NPP vendor. All case study countries in this situation also established cooperative arrangements with the IAEA to support capacity building, education and training, for instance:

- In Belarus, the number of workers for two nuclear power units was estimated to be 2321 people. The training programme includes a basic (5–5.5 years) training in national universities for personnel with higher education for NPP operation, and including practical training abroad, a special (0.5–3 years) training of personnel with working experience at power facilities in Belarus and foreign universities, practical training at existing nuclear facilities abroad, and individual training programmes in the study and training centre in the NPP [55].

  The contract between Rosatom and the Belarusian NPP provides for training of operations personnel to be carried out in several stages. During the first stage an enlarged schedule of HRD for the Belarusian NPP was developed based on the construction schedule, for two years after the signing of the contract. In the second stage, an educational centre was put into operation and staffed with instructor staff, just before the start of operation of the first nuclear power unit. The third stage provides training of operating personnel for the second nuclear power unit of the Belarusian NPP and the development of further human resources management strategies, just before the start of operation of the second nuclear power unit.

- In Egypt, NPPA conducted an integrated study of HR needs which resulted in a staffing plan for the El Dabaa NPP organization. The engineering, procurement and construction (EPC) contract provides on the job training in the areas of design, construction, commissioning, operations and maintenance. Also, as part of the EPC contract, comprehensive training systems including a training centre will be provided by the vendor.

  NPPA reviewed the training outlined under the EPC contract, together with the training provided by its consultant, and incorporated additional training requirements in other contracts. NPPA developed with its consultant comprehensive training programme that includes several elements, such as soft skills, Russian and English language, and project management. In addition, specialized courses are provided in the areas of mechanical, electrical, instrumentation and control, human resources and contract management and negotiations.

- In the UAE, a nuclear operations organizational structure was developed to estimate needed capabilities and staffing. ENEC and its subsidiaries (Nawah and Barakah One Company) worked with local universities to support the recruitment of a qualified UAE workforce for the operation of the Barakah NPP. Personnel for technician positions in Nawah are trained through a Higher Diploma in Nuclear Technology programme set up through a joint venture with Abu Dhabi Polytechnic. In addition, there is an outreach programme to schools to encourage students to study science and advise them of career possibilities at ENEC and its subsidiaries. The prime contractor, Korea Electric Power Corporation (KEPCO), is also responsible for providing qualified individuals to support NPP operations positions.

  Nawah has its own department responsible for recruitment and training of personnel. Nawah developed its own training and qualification programme, as the Korean training model was not fully applicable. It is conducted via a fully operational Technical Training Centre with fully equipped workshop facilities. A modern simulator training centre equipped with two full scope simulators, is available at the plant site.

- In Uzbekistan, a detailed organizational structure was developed for the construction and operation stages in consultation with the vendor. By the end of Phase 3, it is planned that there will be 1867 staff members in the operating organization. Approximately 40% of that number are to be trained under the EPC contract. Staff for the remaining positions, which are less critical for safety and include non-technical positions, are to be hired and trained by Uzatom.

Regarding the HR needs of the regulatory body, in all case study countries the regulatory body participated in the national HR programme while implementing a recruitment and training strategy specific to the needs of its own organization. Some countries' prior experience in regulating research reactors and other nuclear installations facilitated the analysis of HR needs as well as making available personnel with relevant skills for the nuclear programme. Several countries referred to the IAEA Safety Standards Series No. 79, Systematic Assessment of Regulatory Competence Needs (SARCoN) framework for competency management as well as receiving services from the IAEA for training and development of regulatory staff. Some regulatory bodies also received support from the regulatory bodies in the vendor country or from third countries, and/or contracted TSOs to provide external expert support to augment their in-house expertise [56].

- In Belarus, IAEA peer reviews found the early staffing level of Gosatomnadzor and its TSO, JPINR-Sosny to be inadequate to support the workload imposed by the NPP licensing schedule [41]. The government responded to this finding by increasing the budgeted staff allocation of Gosatomnadzor from 32 to 89 positions. A concerted effort to recruit and train the needed additional staff followed with the support of the IAEA and the EU to create a competent and effective state regulatory body for nuclear and radiation safety. For training its staff, Gosatomnadzor used a variety of tools including national training programmes and international instruments. The fact that 81 regulatory employees studied abroad in the framework of IAEA and EU international cooperation projects was recognized as an area of good performance at the Seventh Review Meeting of the Convention on Nuclear Safety in 2017.

  A Scientific and Technical Centre for Nuclear and Radiation Safety (STC NRS) was established by Presidential Decree in 2017 as the State's TSO with responsibility for coordinating the various other State organizations that provide scientific and technical support to Gosatomnadzor [50].

- In Egypt, past experience in licensing and inspecting the existing research reactors led to the availability of relevant skills for the nuclear power programme. ENRRA acquired human capabilities in the area of nuclear installation safety with specialists who have experience of ENRRA activities or have worked at nuclear facilities. A second step was to complete the organization with freshly graduated officers. ENRRA has also contracted with external expert organizations for technical support in the following areas:

    - Development of the regulatory framework.
    - Development of an IMS and electronic management system (EMS).
    - HRD and capacity building.
    - NPP authorization process.

- Planning and implementation of ENRRA regulatory inspection and structures, systems and components (SSCs) conformity assessment.
- Development of tools for emergency coordination in the situation and analytical center.

- In Ghana, the Atomic Energy Commission in collaboration with the University of Ghana, and with support from the IAEA, established the School of Nuclear and Allied Sciences (SNAS) in September 2006 to provide education in nuclear science and technology at the graduate level. The school is providing a pool of knowledgeable personnel for involvement in regulatory oversight of the nuclear power programme. The NRA has developed a training programme with four levels of competence as recommended in the IAEA SARCoN approach. Training activities have been organized with support from IAEA and a variety of international sponsors including the United States of America and the EU. The conversion of the Ghana Research Reactor from high enriched uranium (HEU) fuel to low enriched uranium (LEU) fuel provided an opportunity to demonstrate and practise regulatory oversight of the modification and refuelling of the reactor. A team has been constituted to evaluate the current capacities of the NRA and recommend areas where external support will be needed to support regulatory activities in Ghana.

- Nigeria's reported activity focuses on the NNRA, evaluating the current capacity, identifying gaps in meeting necessary skills, and developing strategies to close the gaps. NNRA has developed a framework to manage the training, development, and maintenance of its staff competence to regulate Nigeria's first NPP based on the IAEA SARCoN model.

- In the UAE, FANR has made significant efforts in recruiting a core team of national and international experienced staff to meet its HR needs. As noted above, FANR employed a 'two track' HR strategy by recruiting international experienced staff to support its work in the early phases of the programme. FANR also contracted international TSOs to augment its in-house resources, to provide specialist expertise, and to help manage workload peaks, for instance during licensing reviews. External experts also provided early support for management system development, drafting regulations and guides, and training.

FANR's human resource strategy for long term sustainability focuses on developing Emiratis to assume positions of increasing responsibility while maintaining an appropriate cadre of international experts. FANR has established a competency framework for organizational positions based on the IAEA SARCoN recommendations. Programmes for development of local talent include scholarships for UAE nationals to pursue a bachelor's or a master's degree in a nuclear-related discipline at universities abroad, an employee development programme designed to equip employees with the necessary knowledge and skills needed to perform their roles, and leadership development based on training courses conducted in the UAE and abroad. The IAEA technical cooperation programme between the IAEA and the UAE provided numerous resources for capacity building and training. UAE staff were also provided with training opportunities in the regulatory body of the Republic of Korea — the vendor country — under an intergovernmental cooperation agreement. FANR today is a multicultural organization with more than 28 different nationalities across the various departments. Approximately 242 employees perform the regulatory functions. Over 70% of the staff are Emirati and over 40% are women, more information is available in FANR 2021 Annual Report [57]

- In Uzbekistan, complementing the national strategy for HRD, the government through a Presidential Resolution in December 2018 has approved the staffing schedule for the Department of Nuclear and Radiation Safety for the period from 2018 to 2028. The State Committee on Industrial Safety of the Republic of Uzbekistan (SCIS) is further developing its competence through a programme of cooperation with the vendor country regulator, the Federal Service for Environmental, Technological and Nuclear Supervision of the Russian Federation (Rostekhnadzor). It has concluded following a cooperation agreement and memorandum of understanding with Rostekhnadzor on cooperation in the field of regulation of industrial, nuclear and radiation safety in the field of atomic energy use for peaceful purposes, along with several other international cooperation agreements.

4.4.4. Challenges faced and lessons learned

For all the case study countries, acquiring the necessary skilled personnel to perform functions when needed at different phases in a new nuclear power programme is a major challenge for all the main organizations involved, namely the NEPIO, the regulatory body and the owner/operator.

For some organizations, regulatory bodies in particular, obtaining a sufficient staffing budget from the central government has been difficult.

The issue of training and development for both the regulatory body and the operating organization has to be approached as a major challenge, especially given that the process of professional development must be carried out in parallel with the performance of employees' functions. It is necessary to establish training programmes considering both the organization as a whole and individual employee. This is a significant and complicated undertaking for HR managers.

Further challenges include the rules and regulations of recruitment and employment in government organizations which in some cases can influence the ability to hire non-nationals, to offer competitive salaries for specialized skills, and to retain qualified staff. The retention of staff after having received training is also reportedly an issue for some organizations.

Some of the approaches that have proved successful among the case study countries are as follows:

- Clear government policy and leadership from the outset of the programme.

- Detailed planning of HR needs and their timing in the key organizations.

- Active coordination of HRD in the different organizations, especially between the utility, the regulatory body, and the educational and training community.

- International cooperation including bilateral cooperation with the vendor country and cooperation with the IAEA in capacity building and training.

- Employment of external TSOs to help manage workload peaks, to provide specialist expertise when needed for specific tasks, and to assist with organizational development.

APPENDIX

CASE STUDY TEMPLATE

COUNTRY A – NUCLEAR COUNTRY

1. NUCLEAR POWER PROGRAM AND CURRENT STATUS

[Click here and type the body of your report]

Points to be considered:
- Nuclear Program
- Status of nuclear power plant
- Project phase and timeline

2. ROLE AND RESPONSIBILITIES OF THE GOVERNMENT AND THE REGULATORY BODY IN PHASE 1 OF THE NUCLEAR POWER PROGRAMME

2.1 National policy and strategy for safety

[Click here and type the body of your report]

Points to be considered:
- Long-term commitment to safety
- Decision making process and stakeholder engagement.
- Fundamental safety objectives and principles
- Provision for human and financial resources.
- Promotion of leadership, management for safety, and safety culture
- Challenges and applied solutions.

2.2 Governmental and legal framework for safety

[Click here and type the body of your report]

Points to be considered:
- Legally binding framework for safety
- Elements to be covered by the nuclear law.
- Authority and responsibilities for organizations (regulator, operator, TSO, research, etc)
- The types of facilities and activities, the type of authorization and the graded approach
- Provisions for regulatory function and the regulatory decision-making process.
- Interfaces with the framework for safety
- Challenges and applied solutions.

2.3. Regulatory infrastructure for safety

[Click here and type the body of your report]

Points to be considered:
- Time frame or history for the establishment of the regulatory infrastructure
- Experiences acquired from regulating nuclear facilities other than NPPs, If applicable
- Authorities, responsibilities, regulatory functions, and organization change, If applicable.
- Human and financial resources
- Functional separation and effective independence
- Challenges and applied solutions.

2.4. Other governmental considerations – global nuclear safety regime

[Click here and type the body of your report]

Points to be considered:
- Participation in the global nuclear safety regime
- International conventions.
- Use of IAEA Safety Standards, and application of the use of internationally harmonized safety requirements, guides and practices.
- International peer reviews.
- Multilateral and bilateral cooperation and assistance programmes.
- Plans for operating and regulatory experience, and use of lessons learned.
- Challenges and applied solutions

3. KEY PRIORITIES FOR THE GOVERNMENT AND THE REGULATORY BODY IN PHASE 1

3.1. Site Survey
[Click here and type the body of your report]

Points to be considered:
- Overview on site survey process
- Criteria related to the acceptability, comparison and ranking of sites.
- Responsibility of site survey at a regional scale (screening and comparison) and identify of candidate sites
- Process of data gathering and analysis results for later phase
- Regulatory arrangements for safety requirements for siting and site evaluations
- Challenges and applied solutions.

3.2. Consultation with interested parties
[Click here and type the body of your report]

Points to be considered:
- Measures for Transparency and openness

- The process of building trust and public communication
- Consultation with interested parties
- Challenges and applied solutions.

3.3. Leadership and management for safety

[Click here and type the body of your report]

Points to be considered:
- leadership for safety and identification of senior managers
- Integration of safety into the management system (goals, strategies and plans)
- Plans for the Key elements of the management system (organization structure, responsibilities, accountabilities, Job descriptions, processes mapping, procedures, etc)
- The process for fostering safety culture
- Plans for monitoring, assessment, and review of the management system
- Challenges and applied solutions.

3.4. Development of Human Resources

[Click here and type the body of your report]

Points to be considered:
- HR planning and the process of determination of needed competences, and resources
- Recruitment process and staffing plans
- Measures to attract and retain qualified staff
- Sources for human resources development (internal and/or external)
- Training programmes and knowledge management
- Use of TSO/External Support Organizations
- Use of IAEA resources/tools were used (e.g., TC & SARCON)
- Challenges and applied solutions.

4. Lesson Learned and Conclusion

[Click here and type the body of your report]

Points to be considered:

- Summary of the generic lessons learned.
- Conclusion.

REFERENCES

[1]   EUROPEAN ATOMIC ENERGY COMMUNITY, FOOD AND AGRICULTURE ORGANIZATION OF THE UNITED NATIONS, INTERNATIONAL ATOMIC ENERGY AGENCY, INTERNATIONAL LABOUR ORGANIZATION, INTERNATIONAL MARITIME ORGANIZATION, OECD NUCLEAR ENERGY AGENCY, PAN AMERICAN HEALTH ORGANIZATION, UNITED NATIONS ENVIRONMENT PROGRAMME, WORLD HEALTH ORGANIZATION, Fundamental Safety Principles, IAEA Safety Standards Series No. SF-1, IAEA, Vienna (2006), https://doi.org/10.61092/iaea.hmxn-vw0a

[2]   INTERNATIONAL ATOMIC ENERGY AGENCY, Governmental, Legal and Regulatory Framework for Safety, IAEA Safety Standards Series No. GSR Part 1 (Rev. 1), IAEA, Vienna (2016).

[3]   INTERNATIONAL ATOMIC ENERGY AGENCY, Establishing the Safety Infrastructure for a Nuclear Power Programme, Safety Standards Series No. SSG-16 (Rev. 1), IAEA, Vienna (2020).

[4]   INTERNATIONAL ATOMIC ENERGY AGENCY, Milestones in the Development of a National Infrastructure for Nuclear Power, IAEA Nuclear Energy Series No. NG-G-3.1 (Rev. 2), IAEA, Vienna (2024), https://doi.org/10.61092/iaea.zjau-e8cs

[5]   INTERNATIONAL ATOMIC ENERGY AGENCY, Generic RoadMap for Establishing the Nuclear Safety Infrastructure for a First Nuclear Reactor (in preparation).

[6]   INTERNATIONAL ATOMIC ENERGY AGENCY, Analysis of the Integrated Regulatory Review Service (IRRS Missions Conducted from 2018 to 2022 to Member States of the European Union, IAEA, Vienna (2023).

[7]   INTERNATIONAL ATOMIC ENERGY AGENCY, Integrated Nuclear Infrastructure Review (INIR): Ten Years of Lessons Learned, IAEA-TECDOC-1947, IAEA, Vienna (2021).

[8]   INTERNATIONAL ATOMIC ENERGY AGENCY, Establishing the Nuclear Security Infrastructure for a Nuclear Power Programme, IAEA Nuclear Security Series No. 19, IAEA, Vienna (2013).

[9]   INTERNATIONAL ATOMIC ENERGY AGENCY, Safeguards Implementation Practices Guide on Establishing and Maintaining State Safeguards Infrastructure, IAEA Services Series No. 31, IAEA, Vienna (2018).

[10]   INTERNATIONAL ATOMIC ENERGY AGENCY, Country Nuclear Power Profiles. Country Nuclear Power Profiles | IAEA

[11]   Convention on Nuclear Safety, INFCIRC/449, IAEA, Vienna (1994).

[12]   Joint Convention on the Safety of Spent Fuel Management and on the Safety of Radioactive Waste Management, INFCIRC/546, IAEA, Vienna (1997).

[13]   INTERNATIONAL ATOMIC ENERGY AGENCY, Power Reactor Information System Database, IAEA. https://pris.iaea.org/pris/home.aspx

[14]   INTERNATIONAL ATOMIC ENERGY AGENCY, Mission Report on the Integrated Nuclear Infrastructure Review (INIR) - PHASE 2 (2019). https://www.iaea.org/sites/default/files/documents/review-missions/inir2-egypt.pdf

[15]   INTERNATIONAL ATOMIC ENERGY AGENCY, The Republic of Ghana 8th and 9th review meeting of the convention on nuclear safety (2022). https://www.iaea.org/sites/default/files/24/02/cns_report_2022_ghana.pdf

[16]   UNITED ARAB EMIRATES, Policy of the United Arab Emirates on the Evaluation and Potential Development of Peaceful Nuclear Energy", Abu Dhabi (2008). https://www.enec.gov.ae/doc/uae-peaceful-nuclear-energy-policy-5722278a2952f.pdf

[17]   REPUBLIC OF BELARUS, RESOLUTION OF THE CABINET OF MINISTERS OF THE REPUBLIC OF UZBEKISTAN of February 1, 2019 No. 75 (2019).

[18]   INTERNATIONAL ATOMIC ENERGY AGENCY, Experiences of Member States in Building a Regulatory Framework for the Oversight of New Nuclear Power Plants: Country Case Studies, IAEA-TECDOC-1948, IAEA, Vienna (2021).

[19]   THE ARAB REPUBLIC of EGYPT, Law No. 7 of the Year 2010 On the Enactment of the Law on Regulation of Nuclear and Radiation Activities, Official Journal, Issue No. 12 (Bis-A), 30 March 2010, Cairo (2010). https://enrra.org/publication/law-regulating-nuclear-radiological-activities/

[20]   THE ARAB REPUBLIC of EGYPT, Law No. 210 of 2017 Amending Some Provisions of Law No. 13 of 1976 on the Establishment of the Nuclear Power Plants Authority for Generating Electricity, Presidency of the Arab Republic of Egypt, Cairo (2017).

[21]   INTERNATIONAL ATOMIC ENERGY AGENCY, Integrated Regulatory Review Service (IRRS) Mission to the United Arab Emirates (2011). https://www.iaea.org/sites/default/files/documents/review missions/irrs_mission_to_uae_dec_2011.pdf

[22]   NTERNATIONAL ATOMIC ENERGY AGENCY, Mission Report on the Integrated Nuclear Infrastructure Review (INIR) – PHASE 3 (2018). https://www.iaea.org/sites/default/files/documents/review-missions/2018-uae-inir-phase-3-report.pdf

[23]   UNITED ARAB EMIRATES, UAE NATIONAL REPORT For the Joint 8th and 9th Review Meetings of the CONVENTION ON NUCLEAR SAFETY (2023). https://www.iaea.org/sites/default/files/24/02/cns_uae_national_report_2022.pdf

[24]   Nigeria, Convention on Nuclear Safety Nigerian National Report for the 8 TH Review Meeting (2019) https://www.iaea.org/sites/default/files/national_report_of_nigeria_for_the_8th_review_meeting.pdf

[25]   REPUBLIC OF BELARUS, Decree of the President of the Republic of Belarus "On Several Measures Aimed at the Nuclear Power Plant Construction" No.565 of November 12, 2007, Belarus (2007).

[26]   INTERNATIONAL NUCLEAR SAFETY GROUP, Strengthening the Global Nuclear Safety Regime, INSAG Series No. 21, IAEA, Vienna (2006).

[27]   Convention on Early Notification of a Nuclear Accident, INFCIRC/335, IAEA, Vienna (1986).

[28]   Convention on Assistance in the Case of a Nuclear Accident or Radiological Emergency, INFCIRC/336, IAEA, Vienna (1986).

[29]   The Convention on the Physical Protection of Nuclear Material, INFCIRC/274/Rev. 1, IAEA, Vienna (1980).

[30]   Amendment to the Convention on the Physical Protection of Nuclear Material, INFCIRC/274/Rev. 1/Mod. 1, IAEA, Vienna (2005).

[31]   Vienna Convention on Civil Liability for Nuclear Damage, INFCIRC/500, IAEA, Vienna (1963).

[32]   Protocol to Amend the Vienna Convention on Civil Liability for Nuclear Damage, INFCIRC/566, IAEA, Vienna (1997).

[33]   Convention on Supplementary Compensation for Nuclear Damage, INFCIRC/567, IAEA, Vienna (1997).

[34]   Joint Protocol Relating to the Application of the Vienna Convention and the Paris Convention INFCIRC/402, IAEA, Vienna (1988).

[35]   Treaty on the Non-Proliferation of Nuclear Weapons, INFCIRC/140, IAEA, Vienna (1986).

[36]   Additional Protocol to CSA, INFCIRC/540, IAEA, Vienna (1997).

[37]   Revised Supplementary Agreement Concerning the Provision of Technical Assistance by the IAEA, INFCIRC/267, IAEA, Vienna (1997).

[38]   INTERNATIONAL ATOMIC ENERGY AGENCY, National Report of the Republic of Belarus on The Implementation of The Convention on Nuclear Safety (2019). https://www.iaea.org/sites/default/files/24/02/cns_belarus_national_report_2022_9th _rm.pdf

[39]   INTERNATIONAL ATOMIC ENERGY AGENCY, Site Evaluation for Nuclear Installations, IAEA Safety Standards Series No. SSR-1, IAEA, Vienna (2019).

[40]   INTERNATIONAL ATOMIC ENERGY AGENCY, Site Survey and Site Selection for Nuclear Installations, IAEA Safety Standards Series No. SSG-35, IAEA, Vienna (2015).

[41]   INTERNATIONAL ATOMIC ENERGY AGENCY, Mission Report on the Integrated Nuclear Infrastructure Review Mission Phases 1 & 2 to Belarus (2012). https://www.iaea.org/sites/default/files/documents/review-missions/2013-01-23_approved_inir_report_belarus.pdf

[42]   INTERNATIONAL ATOMIC ENERGY AGENCY, Mission Report on the Integrated Nuclear Infrastructure Review Mission Phase 1 to Ghana (2017). https://www.iaea.org/sites/default/files/documents/review-missions/inir-mission-to-ghana-january-2017.pdf

[43]   NIGERIAN NUCLEAR REGULATORY AUTHORITY, Regulation on Licensing of Site for NPP, https://nnra.gov.ng/nnra/resources?category=nnra_regulations

[44]   INTERNATIONAL ATOMIC ENERGY AGENCY, Mission Report on the Integrated Nuclear Infrastructure Review Mission Phase 2 to Uzbekistan (2021). https://www.iaea.org/sites/default/files/documents/review-missions/inir2-uzbekistan-030621.pdf

[45]   INTERNATIONAL ATOMIC ENERGY AGENCY, Communication and Consultation with Interested Parties by the Regulatory Body, IAEA Safety Standards Series No. GSG-6, IAEA, Vienna (2017).

[46]   INTERNATIONAL ATOMIC ENERGY AGENCY, Stakeholder Engagement in Nuclear Programmes, IAEA Nuclear Energy Series No. NG-G-5.1, IAEA, Vienna (2021).

[47]   INTERNATIONAL ATOMIC ENERGY AGENCY, Leadership and Management for Safety, IAEA Safety Standards Series No. GSR Part 2, IAEA, Vienna (2016), https://doi.org/10.61092/iaea.cq1k-j5z3

[48]   INTERNATIONAL ATOMIC ENERGY AGENCY, Report of the INIR Phase 3 Mission to Belarus (2020), Issue 3: Management. https://www.iaea.org/sites/default/files/documents/review-missions/inir-3-mission-belarus-040320.pdf

[49]   INTERNATIONAL ATOMIC ENERGY AGENCY, Report of the INIR Phase 3 Mission to the UAE, 24 June–1 July 2018, Issue 3: Management More information is available at Report of the INIR Phase 3 Mission to the UAE (2018). https://www.iaea.org/sites/default/files/documents/review-missions/2018-uae-inir-phase-3-report.pdf

[50]   INTERNATIONAL ATOMIC ENERGY AGENCY, Report of the Integrated Regulatory Review Service (IRRS) Follow-Up Mission to Belarus (2021). https://www.iaea.org/sites/default/files/documents/review-missions/irrs_belarus_follow-up.pdf

[51]   INTERNATIONAL ORGANIZATION FOR STANDARDIZATION, Quality Management Systems: Requirements, ISO 9001:2015, ISO, Geneva (2015).

[52]   INTERNATIONAL ATOMIC ENERGY AGENCY, Organization, Management and Staffing of the Regulatory Body for Safety, IAEA Safety Standards Series No. GSG-12, IAEA, Vienna (2018).

[53]   INTERNATIONAL ATOMIC ENERGY AGENCY, Human Resource Management for New Nuclear Power Programmes, IAEA Nuclear Energy Series No. NG-T-3.10 (Rev. 1), IAEA, Vienna (2022).

[54]   INTERNATIONAL ATOMIC ENERGY AGENCY, Report of the Phase 1 Follow-Up INIR Mission to Ghana (2019), Issue 10: Human Resources Development. https://www.iaea.org/sites/default/files/documents/review-missions/inir-mission-to-ghana-january-2019.pdf

[55]   REPUBLIC OF BELARUS, National Report of the Belarus on the Implementation of the Convention ON Nuclear Safety (2016). https://www.iaea.org/sites/default/files/belarus_nr-7th-en.pdf

[56]   INTERNATIONAL ATOMIC ENERGY AGENCY, Managing Regulatory Body Competence, Safety Reports Series No. 79, IAEA, Vienna (2014).

[57]   FEDERAL AUTHORITY FOR NUCLEAR REGULATION, FANR 2021 Annual Report, FANR, Abu Dhabi (2021).

ANNEXES

The annexes present case studies from six Member States showcasing their experiences in developing regulatory frameworks for new or expanding nuclear power programmes. The annexes are authored by experts from the national regulatory bodies of the respective countries, and do not necessarily reflect the exact views of the IAEA.

ANNEX I.
CASE STUDY ON BELARUS

I-1.   EXISTING SAFETY INFRASTRUCTURE IN BELARUS PRIOR TO INITIATING A NUCLEAR POWER PROGRAMME

Before taking a decision in principle to embark on nuclear power (January 2008), Belarus's infrastructure for the use of nuclear energy and ionizing radiation sources included:

- Scientific nuclear installations in State Scientific Institution "Joint Institute for Power and Nuclear Research – Sosny" (SSI JIPNR–Sosny).
- More than 20 000 ionizing radiation sources in use in medicine, science, industry and other areas.
- National system of radioactive waste management, one radon-type facility to accept all operational radioactive waste from the whole country, as well as decontamination waste disposal sites, which contained radioactive materials with radioactivity level lower than radioactive waste.
- National system of overcoming the Chernobyl catastrophe consequences including legislation, special State authority for coordination of such activities, system of measures in agriculture and economy, radiation control and monitoring system, system of management of territories with high level of radioactive contamination.
- Strong emergency preparedness and response system with huge practical experience.

Functions of regulation in nuclear and radiation safety area were assigned to the State Committee for Supervision of Industrial and Nuclear Safety (Gospromatomnadzor) – a separate legal entity under the Ministry for Emergency Situations which dealt with industrial, nuclear and radiation safety supervision. The Ministry of Health and Ministry of Natural Resources and Environment participated in regulation of nuclear and radiation safety.

I-2.   STATUS OF NUCLEAR POWER PROGRAMME IN BELARUS

As of March 2023, Belarus has successfully started commercial operation of Belarusian nuclear power plant (NPP) power unit No. 1 (on 10 June 2021). Power unit No. 2 is under commissioning. It is expected that it will be put into commercial operation by the end of 2023. By the moment of putting into commercial operation of power unit No. 1 of the Belarusian NPP all necessary elements of nuclear infrastructure including regulatory infrastructure had been in place. This fact was confirmed by a number of IAEA review missions. Belarus hosted all the review missions recommended to embarking countries, including INIR, IRRS, SEED, Emergency Preparedness Review (EPREV), IAEA State system for accounting and control of nuclear material Advisory Service (ISSAS), International Physical Protection Advisory Service (IPPAS) and pre-Operational Safety Review Team (pre-OSART)[5].

---

[5]Some of the content in this annex has been reproduced from the published reports with permission.

I-3. ROLES AND RESPONSIBILITIES OF THE GOVERNMENT AND THE REGULATORY BODY IN BELARUS IN PHASE 1 OF THE NUCLEAR POWER PROGRAMME

I-3.1. National policy and strategy for safety

I-3.1.1. Implementation

Since the very beginning Belarus was involved into activities with nuclear materials based on IAEA Standards Series No. SF-1, Fundamental Safety Principles [I-1]. Corresponding provisions were implemented in national legislation including legislation relating to the Chernobyl accident and the Law of the Republic of Belarus on Radiation Safety of the Population of 1998. Just after the decision in principle to embark on nuclear power the Law of the Republic of Belarus "On the Use of Atomic Energy" was adopted to complement the national legislation with the corresponding provisions taking into account appearance of a new nuclear facility [I-2]. Our practical experience shows that for Phase 1 and Phase 2 of nuclear power programme development it was enough to clearly indicate national policy and strategy for nuclear and radiation safety. This allowed to smoothly pass the next steps including siting, site evaluation, construction, and commissioning. In 2016, the issue of the Belarusian legal framework was analysed by the IAEA IRRS team [I-3]. The team members recommended that the Government develop a separate document, its national policy for safety that addresses the mechanisms for achieving the fundamental safety objective and applying the fundamental safety principles [I-1] in accordance with a graded approach.

With the time passing the legislation continued to develop. In 2021, in Belarus nuclear legislation was formed as a separate legal area. Two new framework laws were prepared and adopted – the Law on Radiation Safety to replace the law of 1998 and the Law on Safety Regulation in the Use of Atomic Energy to replace the atomic law of 2008 [I–4].

Regarding this, as well as taking into account recent changes in the legal framework, a draft National Policy and Strategy in Nuclear and Radiation Safety was developed as a separate document with global view into the future. This document is viewed as being a core for the whole area of nuclear legislation. Adoption of such a document is expected in 2023.

I-3.1.2. Challenges faced and lessons learned

The terms of the development of a national policy and strategy as a separate document in the field of nuclear and radiation safety depends on the specifics of the country.

The experience of Belarus as regards National Policy and Strategy showed that it can be developed as a separate document much later than Phase 1 and Phase 2. The conditions for such an approach are provisions reflecting IAEA Safety Fundamentals [I-1] in the national legislation at the end of Phase 1 – beginning of Phase 2 which allow to smoothly pass siting, site evaluation and construction of nuclear power units. Moreover, the preparation of the National Policy and Strategy at a later stage has its benefits due to taking into account of own experience. This can be a good background for further safe operation of NPP during its whole lifetime.

I-3.2. Governmental and legal framework for safety

I-3.2.1. Implementation

With the beginning of the nuclear programme the intention of Belarus was to build a legal framework based on the IAEA safety standards and documents that describe steps to develop a legislative framework for safety. In Belarus, the legislative framework is built in the form of a pyramid, where at the top level there are the laws of the Republic of Belarus and decrees of the President, which are equivalent in their legal force to laws. Their provisions are detailed by the resolutions of the Government (the Council of Ministers of the Republic of Belarus). Regulatory legal acts of a technical nature are adopted by the State administration bodies, in particular, norms and rules in the field of nuclear and radiation safety are prepared by the Department for Nuclear and Radiation Safety (Gosatomnadzor) and adopted by the Ministry for Emergency Situations of the Republic of Belarus.

At Phase 1 and the beginning of Phase 2 of nuclear power programme development, one of the challenges was the updating of the legal framework in the field of nuclear and radiation safety, taking into account a new nuclear power plant. The work on the update of the legal framework started by taking into account the pre-existing legal framework.

At the early stages of the development of the legislative framework in the field of nuclear and radiation safety in Belarus, international treaties and sanitary norms and rules of the country were used. An important milestone in the qualitative development of the legislative framework was the adoption in 1998 of the Law "On Radiation Safety" [I–4]. Two laws of the Republic of Belarus were adopted in the 1990s and were dedicated to the status of the territories affected as a result of the Chernobyl disaster and social protection of the population.

In 2004, the Decree of the President of the Republic of Belarus "On Licensing of Certain Types of Activities" was adopted [I–5]. , which also defined the requirements for licensing activities in the field of nuclear and radiation safety. Later, the mentioned licensing framework has been continuously improved in line with the requirements of the IAEA and the best international practices. It was done within Phase 2 and at the end of Phase 3 [I–5].

At the beginning of Phase 2 in 2008 in addition to the already existing Law "On Radiation Safety of the Population" [I–6] , a top-level document was developed, the Law of the Republic of Belarus "On the Use of Atomic Energy" [I-2]. , which consolidated a number of provisions that establish the requirements of the IAEA in national legislation.

The safety approach of Belarus at the end of Phase 1 and the beginning of Phase 2 was outlined in the National Security Concept of the Republic of Belarus, [I–3] which described fundamental principles to be applied by the Government. Specific principles applicable to nuclear and radiation safety [I–7] are included in two main legislative acts of the national regulatory framework: the Law of the Republic of Belarus "On the Use of Atomic Energy" and the Law of the Republic of Belarus "On Radiation Safety of the Population" [I–6] .

Article 3 of each law includes the key principles that govern the application of the act and are generally consistent with SF-1[I-1]. Many elements of Requirement 1 of IAEA Safety Standards Series No. GSR Part 1 (Rev. 1), Governmental, Legal and Regulatory Framework for Safety [2] are included.

A key feature of the development of the legal framework of Belarus was the opportunity of using technical regulatory legal acts of the vendor country — Russian Federation — if there were no Belarusian ones and if they are in compliance with the requirements of the IAEA. This allowed the Belarusian regulatory authority to rationally distribute resources and gradually develop its own legal framework on a planned basis. In this case, two-stage annual planning was used. Plans for development of top-level documents were approved at the government level, and technical regulatory legal acts at the level of Gosatomnadzor. The Belarusian regulatory framework was fully formed by the beginning of commercial operation of the Belarusian NPP Unit 1 (end of Phase 3) in 2021.

Taking into account the mentioned circumstances, it was also only by the end of Phase 3 that a separate branch of nuclear legislation was legally formalized. This was done by Presidential Decree No. 137 of April 5, 2021 "On Regulation of Activities in the Field of Use of Atomic Energy and Ionizing Radiation Sources", which introduced the related changes to the Unified Legal Classifier of the Republic of Belarus [I–7].

Following the principle of continuous improvement of safety regardless of the level reached, work on improving the legislative framework based on practical experience, as well as on new IAEA documents, continues. In 2022, the Law "On Safety Regulation in the Use of Atomic Energy" was adopted, which will enter into force in 2023 and will replace the Law "On the Use of Atomic Energy" [I–7].

I-3.2.2.   *Challenges faced and lessons learned*

The adoption of framework documents at the beginning of Phase 2 and the possibility of using Russian technical normative legal acts made it possible to smoothly update national legislation in nuclear and radiation safety and within this work to fix the established practices.

Challenge: amending the legal framework in nuclear and radiation safety to include nuclear power.

I-3.3. Regulatory infrastructure for safety

I-3.3.1.   *Implementation*

The creation of a regulatory infrastructure for safety in Belarus for the nuclear power programme was based on those infrastructure elements that existed in the country before the decision on the construction of a NPP. The most important step in further actions was the establishment and maintenance of an independent regulatory body in the field of ensuring nuclear and radiation safety. This is what Gosatomnadzor has become: a separate legal entity within the structure of the Ministry for Emergency Situations. It was created in accordance with Presidential Decree No. 565 "On Several Measures Aimed at the Nuclear Power Plant Construction" [I–7].

Subsequently, during Phase 2, other elements of the regulatory safety infrastructure were formed and adapted to the current stage of the nuclear power programme, including in the following areas: supervision, licensing, safety assessment, creation of a technical support system, development of legislation and regulatory requirements, information work with various interested parties.

The 2016 IRRS mission noted that regulatory infrastructure was in place in Belarus and suggested measures for its further improvement. Currently, all elements of the regulatory infrastructure continue to develop in accordance with the principles of permanent safety improvement, taking into account own practical experience and new documents summarizing international experience.

I-3.3.2.   *Challenges faced and lessons learned*

The main challenge is to manage the rapid growth of the regulatory body, build up the competencies of specialists and simultaneously perform the main functions. Other challenges include the organization and conduct of an effective safety assessment for the NPP, increasing the capacity of independent technical support in order to meet future tasks, and establishment of an onsite inspection system.

I-3.4. Other governmental considerations – global nuclear safety regime

I-3.4.1. Implementation

Before its decision to embark on nuclear power, Belarus had been an active participant of the global safety regime. In particular, the legislation in this area was built based on the IAEA safety requirements. Belarus became a contracting party of international treaties and conventions as follows [I–7]:

- Convention on Early Notification of a Nuclear Accident and the Convention on Assistance in the Case of a Nuclear Accident or Radiation Situation (ratified by Decree of the Presidium of the Supreme Council of the Republic of Belarus No. 1216-XI of December 18, 1986) [I–10];

- Convention on the Physical Protection of Nuclear Material (ratified by the Decree of the Presidium of the Supreme Council No. 2381-XII of June 14, 1993)[I–11];

- Treaty on the Non-Proliferation of Nuclear Weapons (ratified by the Supreme Council of the Republic of Belarus No. 2166-XI of February 4, 1993) [I–12];

- Agreement between the Republic of Belarus and International Atomic Energy Agency on the application of safeguards in connection with the Treaty on the Non-Proliferation of Nuclear Weapons of August 31, 1995 [I–13];

- Vienna Convention on Civil Liability for Nuclear Damage (ratified by the Law of the Republic of Belarus No. 76-3 of November 11, 1997) [I–14]

- Convention on Nuclear Safety (ratified by the Decree of the President of the Republic of Belarus No. 430 of September 2, 1998) [I–15]

- Joint Convention on the Safety of Spent Fuel Management and on the Safety of Radioactive Waste Management (ratified by the Law of the Republic of Belarus No. 130 of July 17, 2002)[I–16];

- Protocol to Amend the Vienna Convention on Civil Liability for Nuclear Damage (ratified by the Law of the Republic of Belarus of April 30, 2003, No. 187-3) [I–17]

- International Convention for the Suppression of Acts of Nuclear Terrorism (ratified by the Law of the Republic of Belarus of October 20, 2006, No. 171-3)[I–18].

Later on, Belarus took active efforts regarding global safety regime: invited and hosted IAEA peer review missions, enhanced cooperation with other countries on bilateral and multilateral basis. As of March 2023, Belarus has signed and implemented more than 20 bilateral agreements in nuclear and radiation safety area, takes part in the activities of regional international forums and networks (e.g. RCF, European and Central Asia Safety Network (EuCAS), VVER Regulator's Forum, World Association of Nuclear Operators (WANO)) and initiated creation of a professional regulators' forum within Commonwealth of Independent States (CIS) [I–3], among other things.

I-3.4.2.   Challenges faced and lessons learned

Intensive participation in the global safety regime is the key to the success of nuclear and radiation safety infrastructure development. Sustainable interaction with the organizations of the supplier country at different levels is of vital importance.

Participation in the global safety regime necessarily involves the availability of adequate resources, primarily human resources, in the operating organization, regulatory authority and other interested organizations for the implementation of international cooperation and its intensive development.

I-4.   KEY PRIORITIES FOR THE GOVERNMENT AND THE REGULATORY BODY IN BELARUS IN PHASE 1

I-4.1. Site survey

I-4.1.1.   Implementation

The Law "On the Use of Atomic Energy" provides that the issues of the decision on siting, designing, construction, commissioning, prolongation of an operating period, limitation of operational characteristics and decommissioning of the NPP are by the power of the President and Government of Belarus. The following provisions exist:

- Basic criteria and requirements that regulate the siting of the NPP in including the influence of natural and human-induced processes, occurrences, and factors on environment and population.

- Basic requirements for the structure and range of investigations for selecting the location for siting of the NPP in Belarus.

- Requirements for the development and content of the environmental impact assessment, and the ecological substantiation of safety of the NPP.

To implement the plan of main preparatory works before the NPP construction, comprehensive research of three locations (Krasnopolyanskaya, Kukshinovskaya, Ostrovetskaya) was done. It included:

- The analysis of archive data about radiation-chemical contamination of the natural environment from stationary observation stations.

- Hydrometeorological, aerological, and geology-geophysical research.

The analysis of aerological conditions in the areas of potential NPP siting locations revealed that the competitive locations are in approximately equal hydrometeorological conditions. Climatic conditions do not impede the siting of the NPP. The geological structure of alternative locations and their areas, geophysical fields and subsurface structure of the Earth crust were investigated, fractures and tectonic active structures were defined, seismicity and the seismic conditions were investigated as well.

In December 2008, according to results of investigations, the State Commission defined the Ostrovetskaya location as a priority, because this location has the best geological features, has acceptable characteristics and is fully adequate for the NPP siting requirements. Krasnopolyanskaya and Kukshinovskaya locations were defined as reserve locations.

I-4.1.2.   *Challenges faced and lessons learned*

The survey of potential sites for an NPP is a critical step and is carried out at the beginning of implementation of the nuclear power programme. To successfully complete this task, it is necessary to have organizations with a sufficient level of competence in a number of areas, such as hydrometeorology, aerology and geology-geophysics. The presence in the country of such organizations is a significant advantage in terms of the quality and timing of these works and the timing. In Belarus, such organizations were available in the structure of the National Academy of Sciences.

I-4.2. Consultation with interested parties

I-4.2.1.   *Implementation*

Since the approach to the nuclear power programme from the very beginning was that it was a state project, and not a private one, interaction with interested parties in Phase 1 and Phase 2 took place within the framework of the relationship between government authorities and State organizations based on the established long-term practice. This approach was used while justifying the need to include nuclear energy in the country's energy balance and preparing important decisions at the level of the Head of State and the Security Council of the Republic of Belarus.

The country's public was regularly informed about the preparation and subsequent decision-making on the nuclear power programme by the State media. In 2007, Presidential Decree No. 565 "On Several Measures Aimed at the Nuclear Power Plant Construction" [I-9] was adopted, which identified key participants, including newly created ones, namely the following:

- Department for Nuclear and Radiation Safety, which had to be created within the structure of the Ministry for Emergency Situations as a regulatory body.

- Department of Nuclear Energy, which had to be created within the structure of the Ministry of Energy as a supervisor of the promotion of nuclear energy.

- State Enterprise "Directorate for the Construction of a Nuclear Power Plant" (SE "DSAE"), which had to be created to carry out the functions of a customer to perform a set of preparatory and design and survey work on the construction of an NPP and subsequently an operator. In 2014, this state enterprise was transformed into the State Enterprise "Belarusian NPP".

- State Scientific Institution "Joint Institute for Power and Nuclear Research – Sosny" (SSI "JIPNR–Sosny") was defined as an organization that provides scientific support for the construction of the NPP.

After the adoption of the Decree, it was these government bodies and organizations that became key in the context of further implementation of the nuclear power programme. Communication between these interested parties, as well as their communication with others, was carried out within the framework of an established system of interaction in Belarus.

In 2008, with the adoption of the Law "On the Use of Atomic Energy", the role and functions of the participants in the first nuclear power programme were clarified. It was determined that the state administration in the field of the use of nuclear energy in accordance with this Law and other legislative acts is carried out by the Ministry of Energy, the Ministry for Emergency Situations, as well as other government bodies and other State organizations authorized by the President of the Republic of Belarus.

The Ministry for Emergency Situations, the Ministry of Natural Resources and Environmental Protection, the Ministry of, the Ministry of Internal Affairs, and the State Security Committee (hereinafter referred to as the State bodies for regulating safety in the use of nuclear energy) are authorized public administration bodies that carry out State regulation of activities to ensure safety in the use of nuclear energy [I–1].

With the further implementation of Phase 2, two coordination mechanisms were created by the resolutions of the Government (Council of Ministers of the Republic of Belarus):

- Inter-ministerial Commission for the Coordination of the Plan of Basic Organizational Measures for the Construction of a Nuclear Power Plant in the Republic of Belarus and control over its implementation, headed by the Vice Prime Minister of Belarus.
- Working group for coordination the implementation of state control (supervision) for the construction of an NPP, headed by the Deputy Minister for Emergency Situations.

Subsequently, the existence of these coordination mechanisms was recognized as good practice by the results of the IAEA IRRS mission for a comprehensive assessment of the regulatory infrastructure in 2016.

Understanding the importance of a systematic approach to the organization of interaction with interested parties, in 2013 Gosatomnadzor adopted the Information and Communication Strategy [I–7] as a separate document, which was subsequently updated twice in 2016 and 2021.

The objective of the implementation of the Strategy is the establishment, maintenance and development of the constructive and creative dialogue between the Gosatomnadzor and its members on the one hand, public authorities, organizations, experts, media and public (Belarusian and international), on the other hand, to achieve a common understanding of the issues of nuclear and radiation safety, the role and functions of the regulatory body, and the development of a safety culture in the nuclear field.

Main tasks of the Strategy:

Achieve sustainable awareness by various target audiences of the Ministry for Emergency Situations as an authorized State body and the regulator in the field of nuclear and radiation safety, and of Gosatomnadzor as the Department of the Ministry for Emergency Situations delegated with powers of State supervision in the field of nuclear and radiation safety and control over the implementation of the relevant legislation.

Strengthen the credibility of the State regulation of nuclear and radiation safety in Belarus as a whole, as well as of Gosatomnadzor, based on public demonstration of practical actions of the latter to implement powers granted to it in terms of State supervision in the field of nuclear and radiation safety and control over the implementation of the relevant legislation.

Develop adequate representation for domestic and foreign audience (professionals and public) on up-to-date assurance of nuclear and radiation safety in Belarus.

Develop safety culture as an essential professional attribute in the nuclear field.

The main interested parties are the following:

- Professionals (users engaged in the operation of nuclear facilities and sources of ionizing radiation).

- Experts and governmental agents (of Belarus, and of foreign and international organizations) involved in the development of nuclear power, assurance of nuclear and radiation safety (including in the personnel training, regulatory authority technical support), other areas which may be associated with activities on assurance of nuclear and radiation safety [I–3].

- Professionals who wish to link their professional activity to nuclear power, nuclear and radiation safety assurance, related fields (including students of Belarusian higher education institutions which train specialists for nuclear power).

- Members of the public (Belarusian, foreign and international, including those holding responsible positions not related to nuclear and radiation safety assurance) who are interested in and/or concerned about the issues of nuclear and radiation safety assurance, including in the context of NPP construction

- Members of the media are participating in the coverage and public discussion of issues of nuclear and radiation safety, who are the mediators in the dialogue of professionals and public.

For each group of interested parties an information flow scheme was prepared.

Gosatomnadzor in 2012 launched its website, organized intensive interaction with the mass media, and published annual reports on the status of nuclear and radiation safety in Belarus.

One of the most challenging issues within Phase 2 and later was the organization of public discussions and consultations according to the procedures regarding Environmental Impact Assessment Report of the Belarusian NPP. Belarus had to follow official procedures of the Convention on Environmental Impact Assessment in a Transboundary Context [I-7]and in this regard public discussions and consultations were organized in Belarus and abroad.

Gosatomnadzor as the national regulatory authority conducted the public hearings about the important decisions much later after studying best international experience and practice and creation of the corresponding legal base. The first public hearings in the history of the regulatory body were conducted in 2021 before issuing the license for the Belarusian NPP Unit 1 commercial operation.

I-4.2.2. *Challenges faced and lessons learned*

The strong pre-existing system of interaction between governmental authorities and organizations, and the attitude to nuclear power as a State-wide (or national) project helped in running a seamless process of embarking on nuclear power.

I-4.3. Leadership and management for safety

I-4.3.1. *Implementation*

In the case of Belarus, the peculiarity was the fact that from Phase 1, the construction of the Belarusian NPP was considered as a state project. The heads of organizations involved in the development of the nuclear power programme at all levels paid due attention to the practical aspects of its implementation.

During the implementation of Phase 2, management and leadership for safety was strengthened by the creation of the Inter-ministerial Commission for the Coordination of the Plan of Basic Organizational Measures for the Construction of a Nuclear Power Plant in the Republic of Belarus. Leadership and management were also practically implemented through the creation and implementation of quality assurance programmes and integrated management systems (IMSs) for safety purposes, as well as the development of a safety culture in the organizations involved, primarily in the regulatory body Gosatomnadzor and the operating organization State Enterprise "Belarusian NPP". Work in this direction was carried out for a long-time during Phase 2 and Phase 3.

The highest priority principle in ensuring safety has been declared by the management of the State Enterprise "Belarusian NPP" in its Safety Policy.

As part of the preparation for industrial operation, the State Enterprise "Belarusian NPP" elaborated a Plan for the implementation of a system of measures for sustainable independent functioning of the State Enterprise "Belarusian NPP" as an operator. The plan contains measures, including further improvement of approaches to demonstrate leadership in safety issues, the level of safety culture, the efficiency and effectiveness of the IMS, personnel training system, ensuring interaction with WANO structures, international organizations, TSO, etc. [I–3]

To create the necessary conditions for the continuous improvement of the safety culture, both at the organizational level and at individual level, to promote adherence to the safety culture among employees, to maintain an atmosphere of trust and openness in safety issues, an advisory coordinating body was created under the Director General of the operating organization – the Council for Safety Culture of the State Enterprise "Belarusian NPP".

To facilitate the effective functioning of the system for formation and maintenance of the safety culture, executives in charge of safety culture were appointed in the structural units of the Belarusian NPP.

At the most significant (regarding the basis laid for the subsequent safe operation of the facility) stage of the NPP lifetime – construction and commissioning of power units – the permanent supervision mode is enshrined in the Decree of the President of the Republic of Belarus No. 62 of February 16, 2015 "On Ensuring Safety During the Construction of the Belarusian Nuclear Power Plant" [I–19] and for other (except Gosatomnadzor) State bodies (organizations) implementing types of control and supervisory activities in Belarus related to the provision of various aspects of safety. A tool is provided for organizing interaction and coordination of their activities on site – a Working Group involving the top management of such State bodies (State organizations), as well as a separate type of inspections – comprehensive inspections performed with the participation of all or most of these State bodies. This approach was subsequently noted by the IAEA peer review as a good practice with its use recommended in other countries with a nuclear programme [I–3].

In the time of construction and commissioning of the Belarusian NPP, Gosatomnadzor also conducted inspections of manufacturers and suppliers of key equipment important for safety with a visit directly to production facilities.

A requirement of annual safety assessment by the operating organization is established, as well as by organizations involved in work and providing services to the operating organization, including an appropriate report submission to the regulating body. The outcomes of such assessment are the basis for the development and adoption of appropriate regulatory measures by Gosatomnadzor.

In 2021, Gosatomnadzor developed and put into effect the norms and rules for ensuring nuclear and radiation safety "General requirements for the operating organization management systems in order to ensure nuclear and radiation safety" [I-20]. These norms and rules establish general technical requirements for the constituent elements, the procedure for the functioning and improvement of administrative systems of the operating organization in terms of ensuring nuclear and radiation safety at all stages of the life cycle of nuclear energy facilities.

In addition, in 2022 Gosatomnadzor approved the guidelines on nuclear and radiation safety "Ensuring a safety culture at all stages of the life cycle of the Belarusian NPP" [I-21] comprising recommendations on development and maintenance of a safety culture at all stages of the NPP lifetime.

Safety issues are the key issues during regular meetings of the regulatory authority officials and the management of the operating organization.

Safety priority within the regulator's own activities

The high priority of safety is established at the level of the laws of the Republic of Belarus "On the Use of Atomic Energy" [I–2] and "On Radiation Safety"[I–4], and expressed in the strategic documents of Gosatomnadzor (Gosatomnadzor Policy and Strategy approved by the decisions of the Gosatomnadzor Board).

The priority of safety is the basis of the regulatory body's activities within the framework of implementing functions and tasks assigned to, the regulatory body, including:

- Developing norms and rules to ensure nuclear and radiation safety.
- Implementing State supervision in the field of nuclear and radiation safety.
- Preparing decisions on issuing a licence, introducing or rejecting amendments thereto, extending or refusing to extend the validity period, suspension, renewal, termination, cancellation of the licence based on the results of consideration of received documents, conformity assessment and safety examination.
- Issuing permits for the right to implement activities in the field of atomic energy use and for the realization of educational programmes for advanced training of managers and specialists on nuclear and/or radiation safety.
- Licence control.
- Safety assessment.

Gosatomnadzor consistently implements an IMS, one of the tasks of which is to increase the level of safety culture of its own employees. The principles of the Gosatomnadzor safety culture and its attributes were formed with the participation of all Gosatomnadzor employees, enshrined in the Regulation on the IMS of Gosatomnadzor. [I–7]

Gosatomnadzor is continuously developing the safety culture within organization using the outcomes of external audits (the IAEA missions, inspections of Gosatomnadzor activities by the Ministry for Emergency Situations, other competent authorities of Belarus), of the analysis of the Gosatomnadzor activities to provide its efficiency and through detecting non-compliance and elaborating prompt measures for the improvement of activities, taking into account the experience of regulatory bodies in other countries.

A systematic development of competencies of the Gosatomnadzor employees is arranged (including that for the administration of Gosatomnadzor and the Ministry for Emergency Situations). Safety culture issues are included in mandatory training and knowledge inspection of newly accepted employees.

Special events are organized for younger Gosatomnadzor employees (e.g. information sessions with the Head of Gosatomnadzor, professional skills contests, training) to show commitment to the safety culture at all levels.

As part of this activity, the Code of Professional Conduct for Gosatomnadzor Employees was adopted, which represents a set of general principles of professional ethics and basic rules of conduct to guide Gosatomnadzor employees in performing their work duties to implement faithfully, properly, effectively and at a high professional level their professional activities and contribute to strengthening the image of Gosatomnadzor. Familiarization with the Code of newly accepted employees of Gosatomnadzor is mandatory.

In order to define the state of the safety culture in Gosatomnadzor, timely identification of problematic issues, negative trends and positive practices in the field of safety culture to make decisions aimed at developing a safety culture, a self-assessment of safety culture was organized in 2021 based on the Guidelines for Safety Culture Self-Assessment for the Regulatory Body (IAEA Services Series 40, Vienna, September 2019) [I-22]: training of Gosatomnadzor employees was organized, an IAEA expert mission was conducted to provide methodological assistance in preparing for self-assessment, a basis for self-assessment was created (a working group was created, the procedure for conducting was approved, methodological materials were developed), self-assessment was conducted using all methods recommended by the IAEA, and measures to develop a safety culture were developed.

For the collective discussion of critical issues, a Board was established at Gosatomnadzor comprising of the Head of Gosatomnadzor (Chair of the Board), his deputies and other administrative staff of Gosatomnadzor.

To discuss safety issues, Gosatomnadzor, if necessary, initiates consideration of safety issues at meetings of the National Commission of Belarus for Radiation Protection under the Council of Ministers of the Republic of Belarus, at meetings of the Inter-ministerial Commission for the Coordination of the Plan of Basic Organizational Measures for the Construction of a Nuclear Power Plant in the Republic of Belarus and monitoring its implementation, at the Working Group for Coordinating the Supervision of the Belarusian NPP Construction, and at operational meetings at the NPP construction site involving a wide range of specialists from government bodies of public administration, scientific organizations and enterprises engaged in activities in the field of atomic energy and Incident Reporting Systems for Nuclear Installation (IRS) use.

The most relevant issues of nuclear and radiation safety and significant regulatory decisions are highlighted by Gosatomnadzor in the mass media and the official website of Gosatomnadzor.

I-4.3.2. *Challenges faced and lessons learned*

The commitment from the very beginning of the NPP project as a national project has a positive impact on the promotion of leadership and safety culture at various levels in all organizations involved.

Special attention should be paid to management of big and rapid growth taking into account performing the functions and training in parallel.

I-4.4. Development of human resources

I-4.4.1. *Implementation*

A national training system necessary to support nuclear power with highly qualified specialists, as well as to maintain an appropriate level of knowledge for safe, reliable and efficient NPP operation has formed in the country. The training system includes a set of organizational and technical actions of State administrative bodies, higher and secondary specialized educational institutions, industrial and technical schools, and other public institutions.

To provide training in the field of nuclear energy the National Staff Training Programme for nuclear energy was implemented in Belarus in 2008–2020. The Programme was approved by the Council of Ministers of the Republic of Belarus on September 10, 2008, No. 1329 [I–23].

The goal of this Programme is to organize the complex training system ensuring the acquisition of knowledge and skills necessary for NPP construction and safety operation, nuclear and radiation safety, safety for NPP staff, population, and environment. According to the needs, the official decision for workforce training is formed based on State bodies' requests: the scope of training, retraining, advanced training of specialists, scientists of high qualification with specialty and workers is defined annually educational institutions presenting staff training are defined; annual training plans in respective educational institutions are formed.

Within the framework of the State programme [I–3]:

- National institutions for higher educational (Belarusian State Technical University, Belarusian State University of Informatics and Radio Electronics, Belarusian State University, International Sakharov Environmental University) have started student training on eight new professions in the field of nuclear energy, the total amount of training is 220 people per year.
- The training of teachers and scientists of higher educational institutions abroad has been organized.
- Field trips for students to countries with developed nuclear power programmes have been organized.

Specialist training was opened in 2008 in higher and secondary specialized educational institutions of Belarus, in new specialties such as nuclear physics and technology, construction of thermal and nuclear power stations, Steam turbine plants of nuclear power stations, and electronic control systems in nuclear power stations.

The training programme of personnel with higher education for NPP operation includes a basic (5–5.5 years) training in the universities of Belarus, including practical training abroad, a special (0.5–3 years) training of personnel with working experience at power facilities of Belarus in foreign universities, practical training at existing nuclear facilities abroad, individual training programmes in the study and training centre in NPP. Currently the State programme is being updated, with an assessment of needs in human resources and financing.

According to the overriding importance of training issues for the nuclear power programme, an IAEA technical cooperation project on development human resources and infrastructure for the nuclear power in Belarus is in force currently in Belarus. Ministry of Energy and Ministry of Education and the National Academy of Sciences of Belarus are coordinators of this programme. The programme provides consultations on the issues of a training system creation for nuclear energy considering international practices and recommendations of the IAEA. It includes seminars and training sessions, visits of Belarusian scientists and university teachers to training centres of NPP and to scientific research institutions abroad, visits of Belarusian specialists to existing NPPs and those under construction, as well as the development and supply of computer courses for organizations which participate in the project of the NPP construction in Belarus.

A sequence of IAEA and EU technical cooperation project components to create a competent and effective State regulatory body for nuclear and radiation safety, strengthening and modernization in the field of licensing and inspection, were implemented.

Thus, the regulatory authority used a variety of tools for the training of its staff, including national training programmes and international instruments. The fact that 81 regulatory employees studied abroad in the framework of IAEA and EU international cooperation projects and on bilateral basis was recognized as an area of good performance at the Seventh Review Meeting of the Convention on Nuclear Safety in 2017.

According to the Decree of the President of the Republic of Belarus of March 29, 2011, № 124 "Measures to implement international agreements on civil liability for nuclear damage" [I-24] SE "DSAE" was nominated as the operating organization of "Belarusian NPP". The quantity of industrial workers for two nuclear power units was provisioned to be 2321 people.

SE "DSAE" management provided selection, training, access to independent work and maintaining of operational staff qualifications. The "Belarusian NPP" system of selection and operational staff qualification is aimed at the achievement, control, and maintenance of a qualification level necessary to secure "Belarusian NPP" operation in all modes, as well as at the implementation of actions for mitigation of the consequences of accidents when they occur.

The operating organization provided selection, training, retraining and advanced qualification of workers (staff), as well as maintenance of their necessary number throughout the NPP lifetime. Availability of qualified staff in the field of nuclear and radiation safety is a common requirement for a special permit (licence) to carry out activities in the field of the use of nuclear energy and ionizing radiation sources. The specialists of the licence applicant (licence holder) passed training on nuclear and radiation safety issues not later than one month from the date of assignment to a position and periodically according to the requirements of the regulatory legal acts, but not less than once every five years. Education of specialists was carried out in educational institutions (centres), which have licences issued by Gosatomnadzor to implement training, retraining and advanced training of personnel responsible for nuclear and radiation safety, as well as personnel responsible for regulatory control and conformity assessment at the facilities and manufacturers supervised by Gosatomnadzor [I–3].

The basic criteria and principles to ensure nuclear safety at all phases and in all activities were established ,by the requirements for the operating organization, including the selection and training of operating personnel The operating organization undertook necessary measures for the selection and training of personnel, and the creation of an environment in which safety is considered as vitally important issue and the subject of personal responsibility among personnel.

To implement cooperation, the State Atomic Energy Corporation "Rosatom" ensured training of Belarusian specialists by the general contractor, under the agreements (contracts).

According to the General contract for the construction of the NPP in Belarus (signed on July 18, 2012 in Minsk between Open Joint Stock Company "Nizhny Novgorod Engineering Company "Atomenergyproject", which is a managing company of the Closed Joint Stock Company "Atomconstructionexport" (CJSC ACE) and SE "DSAE") the training of the NPP operating personnel was implemented by the Russian Federation and held in Belarus, on building and operating an NPP in the Russian Federation. This work involved several stages.

During the first stage of the training programme – two years after the signing of the contract – an enlarged schedule of training the staff of the "Belarusian NPP" was developed based on the construction schedule. Then, regulatory legal acts regulating the requirements to the operation staff of the "Belarusian NPP" were developed and the selection and training of "Belarusian NPP" operation staff were performed.

At the second stage, before the start of operation of the first nuclear power unit, in 2016 an educational centre was put into operation and staffed with instructor staff according to the schedule for the first nuclear power unit of the "Belarusian NPP".

The third stage, before the start of operation of the second nuclear power unit, provides for the training of operating personnel for the second nuclear power unit of the "Belarusian NPP" and the development of further human resources management strategies, taking into account further development of nuclear energy.

I-4.4.2. *Challenges faced and lessons learned*

The issue of training, both in the regulatory body and in the operating organization, needs to be approached as a major challenge, especially given that the process of professional development needs to be carried out in parallel with the performance of employees' functions.

In the case of Belarus, this challenge was identified by the Regulatory Cooperation Forum (RCF) during the formation of the RCF support programme for Belarus.

A training programme needs to be formed both for the mentioned organizations as a whole and for individual employees. This is a major and complicated undertaking for HR managers.

REFERENCES TO ANNEX I

[I–1]   EUROPEAN ATOMIC ENERGY COMMUNITY, FOOD AND AGRICULTURE ORGANIZATION OF THE UNITED NATIONS, INTERNATIONAL ATOMIC ENERGY AGENCY, INTERNATIONAL LABOUR ORGANIZATION, INTERNATIONAL MARITIME ORGANIZATION, OECD NUCLEAR ENERGY AGENCY, PAN AMERICAN HEALTH ORGANIZATION, UNITED NATIONS ENVIRONMENT PROGRAMME, WORLD HEALTH ORGANIZATION, Fundamental Safety Principles, IAEA Safety Standards Series No. SF-1, IAEA, Vienna (2006), https://doi.org/10.61092/iaea.hmxn-vw0a

[I–2]   REPUBLIC OF BELARUS, Law of the Republic of Belarus of 30.07.2008 No.426-3 (revised on 22.12.2011) "On Nuclear Energy Use", Belarus (2011).

[I–3]   INTERNATIONAL ATOMIC ENERGY AGENCY, Integrated Regulatory Review Service (IRRS) Mission to Belarus (2016). https://www.iaea.org/sites/default/files/documents/review-missions/irrs_report_belarus_october_2016.pdf

[I–4]   REPUBLIC OF BELARUS, Law of the Republic of Belarus No. 198-3 of June 18, 2019 "On Radiation Safety"; Belarus (2019).

[I–5]   INTERNATIONAL ATOMIC ENERGY AGENCY, Experiences of Member States in Building a Regulatory Framework for the Oversight of New Nuclear Power Plants: Country Case Studies, IAEA-TECDOC-1948, IAEA, Vienna (2021)

[I–6]   REPUBLIC OF BELARUS, Law of the Republic of Belarus of January 5, 1998, No.122-3 "On Radiation Safety of Population" (revised on 04.01.2014 No. 106 -3).

[I–7]   NATIONAL REPORT OF THE REPUBLIC OF BELARUS ON THE IMPLEMENTATION OF THE CONVENTION ON NUCLEAR SAFETY (2022).

[I–8]   INTERNATIONAL ATOMIC ENERGY AGENCY, Governmental, Legal and Regulatory Framework for Safety, IAEA Safety Standards Series No. GSR Part 1 (Rev. 1), IAEA, Vienna (2016).

[I–9]   REPUBLIC OF BELARUS, Decree of the President of the Republic of Belarus "On Several Measures Aimed at the Nuclear Power Plant Construction" No.565 of

November 12, 2007, Belarus (2007). https://gosatomnadzor.mchs.gov.by/upload/iblock/523/cns-belarus-national-report-2022-en.pdf

[I–10] Convention on Early Notification of a Nuclear Accident, INFCIRC/335, IAEA, Vienna (1986).

[I–11] Convention on the Physical Protection of Nuclear Material, INFCIRC/274/Rev. 1, IAEA, Vienna (1979).

[I–12] Treaty on the Non-Proliferation of Nuclear Weapons, INFCIRC/140, IAEA, Vienna (1986).

[I–13] Treaty on the Non-Proliferation of Nuclear Weapons, INFCIRC/140, IAEA, Vienna (1986).

[I–14] Vienna Convention on Civil Liability for Nuclear Damage, INFCIRC/500, IAEA, Vienna (1963).

[I–15] Convention on Nuclear Safety, INFCIRC/449, IAEA, Vienna (1994).

[I–16] Joint Convention on the Safety of Spent Fuel Management and on the Safety of Radioactive Waste Management, INFCIRC/546, IAEA, Vienna (1997).

[I–17] Protocol to Amend the Vienna Convention on Civil Liability for Nuclear Damage, INFCIRC/566, IAEA, Vienna (1997).

[I–18] International Convention for the Suppression of Acts of Nuclear Terrorism (ICSANT), UNODC, Vienna (2007)

[I–19] REPUBLIC OF BELARUS, Decree of the President of the Republic of Belarus No. 62 of February 16, 2015 "On Ensuring Safety During the Construction of the Belarusian Nuclear Power Plant", Belarus (2015).

[I–20] REPUBLIC OF BELARUS, Norms and rules for ensuring nuclear and radiation safety "General requirements for the operating organization management systems in order to ensure nuclear and radiation safety", approved by Resolution of the Ministry for Emergency Situations of the Republic of Belarus No. 73 of October 18, 2021, Belarus (2021).

[I–21] REPUBLIC OF BELARUS, Nuclear and Radiation Safety Guidelines "Ensuring a safety culture at all stages of the life cycle of the Belarusian NPP", approved by Gosatomnadzor Order No. 7 of February 04, 2022, Belarus (2022).

[I–22] INTERNATIONAL ATOMIC ENERGY AGENCY, Guidelines for Safety Culture Self-Assessment for the Regulatory Body, IAEA Services Series No. 40, IAEA, Vienna (2019).

[I–23] REPUBLIC OF BELARUS, Resolution of the Council of Ministers of the Republic of Belarus of September 10, 2008, No.1329 "On approving the State Program of Staff Training for the Nuclear Power Industry of the Republic of Belarus for 2008-2020" (revised on 28.11.2013), Belarus (2013).

[I–24] REPUBLIC OF BELARUS, Decree of the President of the Republic of Belarus No. 124 of March 29, 2011 "On measures aimed at implementation of international instruments on civil liability for nuclear damage", Belarus (2011).

ANNEX II.

CASE STUDY ON EGYPT

II-1. EXISTING SAFETY INFRASTRUCTURE PRIOR TO INITIATING A NUCLEAR POWER PROGRAMME IN EGYPT

The Egyptian Atomic Energy Authority (EAEA) is the owner and operator of two research reactors:

- Egypt First Research Reactor (ETRR-1): Tank type 2 MW VVR research reactor is fuelled with well-known Al clad, 10% enriched EK-10 type fuel rods. ETRR-1 reactor went critical for the first time in 1961. The reactor is currently in an extended shutdown since April 2010.
- Egypt Second Research Reactor (ETRR-2): Open pool type research reactor of 22 MW thermal power cooled and moderated by light water with a beryllium reflector and low enriched uranium (LEU) fuel. ETRR-2 reactor went critical for the first time in 1997. The reactor is currently operable and licensed for operation at full power.

The Regulation of Egypt's research reactors during construction and operation contribute significantly to the development of safety infrastructure prior to initiating the nuclear power programme, including the legal and regulatory framework infrastructure [II–1].

The National Centre for Nuclear Safety and Radiation Control (NCNSRC) was established in 1991 as a part of the Egyptian Atomic Energy Authority (EAEA). NCNSRC, as the Egyptian regulatory body, issued a construction permit and fuel loading permit for ETRR-2 and permissions to raise the reactor power up to 22 MW (1998). The preliminary operation license for ETRR-2 was issued (September 1998) by NCNSRC based on Law No. 59 (1960) for the "Uses of Ionization Radiation and Protection Against its Hazards"[II–2], review and assessment of the final safety analysis report (SAR) of the reactor, and the results of commissioning.

The Egyptian regulatory body was involved in pre-operational tests and fuel loading and power rise attendance and conducting regulatory inspections for Egyptian research reactors (established inspection programme in 2004 with the support of IAEA expert mission). The inspection programme is amended by: Regulatory Procedures on Inspection of Nuclear installations in Operating Stage (2016) [II-3].

The Nuclear Law (Law no. 7/2010) established the Egyptian Nuclear and Radiological Regulatory Authority (ENRRA) [II-2] as the independent regulatory authority, defining its main roles and responsibilities, to ensure the safe use of nuclear energy and ionizing radiation. ENRRA is responsible to carry out regulatory control related to nuclear and radiological activities and facilities in Egypt in a way that protects humans, property, and the environment against the hazard of exposure to ionizing radiation.

Egypt's research reactors developed human resources and nuclear education and training. The developed human resources have contributed strongly to understanding the country's needs in order to make knowledgeable decisions regarding nuclear power. Egypt's research reactors also maintained the national safety and security cultures that are difficult to achieve in the absence of active programmes and hands-on experience. Utilization and building upon experience acquired from regulating of research reactors can support beginning a nuclear power programme.

II-2. STATUS OF NUCLEAR POWER PROGRAMME IN EGYPT

In early 2015, Egypt decided to select a strategic partner for the implementation of its first commercial NPP at the El Dabaa site. Following this decision, in November 2015, an intergovernmental agreement (IGA) was signed between the Government of the Russian Federation and the Government of Egypt on Cooperation in Construction and Operation of four units of the NPP (VVER-1200), including fuel supply, used fuel, training and development of regulatory infrastructure.

The IGA was followed by four commercial contracts signed by NPPA with Russian companies in December 2017: (1) Engineering, Procurement and Construction (EPC) Contract; (2) Nuclear Fuel Supply (NFS) Contract; (3) Operation, Support and Maintenance (OSM) Contract; and (4) Spent Nuclear Fuel Treatment (SNFT) Contract [II–1].

In 2019, NPPA received from ENRRA a site approval permit for the El Dabaa site [II–4]. On 28 February 2022, ENNRA issued the site permit of Spent Fuel Facility of El Dabaa NPP.

The set construction permit documents, including preliminary safety analysis report (PSAR) chapters for four units of the NPP (VVER-1200), were completed at the end of 2021 for review and assessment for a construction permit.
The ENRRA Board, on 29 June 2022, granted the construction permit of El Dabba NPP Unit 1. The first concrete of Unit 1 was poured in July 2022, and regulatory activities were scheduled according to the construction plan.

On 31 October 2022, the construction permit of El Dabba NPP Unit 2 was granted. The first concrete of Unit 2 was poured in November 2022, and regulatory activities were scheduled according to the construction plan.

The construction permit for El Dabba NPP Unit 3 was granted on 30 March 2023, and the construction permit for Unit 4 is expected to be issued in the third quarter of 2023.

II-3. ROLES AND RESPONSIBILITIES OF THE GOVERNMENT AND THE REGULATORY BODY IN EGYPT IN PHASE 1 OF THE NUCLEAR POWER PROGRAMME

II-3.1. National policy and strategy for safety

II-3.1.1. Implementation

In 2015 Egypt issued Integrated Sustainable Energy Strategy to 2035. The executive summary of this document provides a justification for the nuclear power programme and considers several scenarios involving a total contribution from nuclear power of 4800 MW by 2030 [II–5].

Law No. 7/2010 (Nuclear Law) establishes the Nuclear and Radioactive Control Authority (known as the Egyptian Nuclear and Radiological Regulatory Authority or ENRRA) as the independent authority responsible for the regulation of safety, security and safeguards [II-2].

Law No. 13/1976, as amended by Law 210/2017, identifies NPPA as the authority responsible for constructing and operating NPPs and states that NPPA will execute nuclear power plant projects considering the latest scientific, technological and safety measures, either by itself or through a third party [II–6].

The NEPIO function is carried out by the Supreme Council and the Coordination Committee. The Supreme Council is headed by the President of the Republic and its members include the Prime Minister and several ministers. The Minister of Electricity and Renewable Energy (MOERE) provides the secretariat for the Supreme Council.

The Supreme Council approves national strategies and plans, promotes international cooperation, and issues decisions to establish or restructure bodies and agencies. The Supreme Council meets once a year or more frequently, if needed.

The Coordination Committee is under the Supreme Council and is headed by the MOERE. Members of the Coordination Committee include staff from relevant ministries, State agencies, and technical staff working to oversee the infrastructure development in different areas. The Coordination Committee meets monthly and as necessary. As an example of how the NEPIO function works, the strategy on spent nuclear fuel, radioactive waste and decommissioning was developed by a working group, reviewed by the Coordination Committee and approved by the Supreme Council. The strategy was then used to finalize the negotiations for the nuclear fuel supply contract with the technology provider. There is a monthly meeting (or more, if needed) between the chairpersons of the Coordination Committee and the Supreme Council to discuss the El Dabaa project's implementation issues. Topics related to safety, security and safeguards were discussed on several occasions. The Supreme Council encourages knowledge transfer through bilateral agreements with countries with experience in nuclear power [II-1].

II-3.1.2.   Challenges faced and lessons learned

The Government of Egypt established a clear process for the coordination of all activities related to the establishment of the national nuclear infrastructure including the communication with all interested parties This challenge was resolved by the Coordination Committee under the Supreme Council and headed by the MOERE provided the coordination of such activities. Members of the Coordination Committee include staff from relevant ministries, State agencies, and technical staff working to oversee the infrastructure development in different areas. The Coordination Committee meets monthly and as necessary.

II-3.2. Governmental and legal framework for safety

II-3.2.1.   Implementation

Legal framework

The experience gained from regulating Egyptian nuclear facilities (other than NPP) helped in identifying all necessary elements of a legal framework for the safety infrastructure, including regulatory functions and responsibilities, and helped in preparation and approval of nuclear law (Law No. 7, 2010) for regulating nuclear and radiological activities and structure and developing the legal framework [I–1].

The Egyptian Government established in March 2010, Law No. 7 for regulating nuclear and radiological activities [II-2]. The Law No. 7 (and its amendment) and its Executive Regulation form the top level of Egypt's legal framework [II–7]. The Nuclear Law aims at setting a legal framework to regulate all nuclear and radiological activities in Egypt in a way that ensures the safety and protection of human beings, property, and the environment against the radioactive hazards. The provisions of this Law apply to the nuclear and radiological activities and exercises in all fields. Law No. 7/2010 for regulating nuclear and radiological activities in Egypt contents [II-2][II–8]:

- Scope and general definitions.
- Adherence to the peaceful use of nuclear energy.
- Independence, responsibilities and functions, and financial resources of the regulatory body.
- Regulatory body board authorities and responsibilities.
- General rules of licensing and permits, inspection, and enforcement.
- Licensees' obligations.
- Nuclear power plant licences.
- Mining and processing facilities.
- Export, import, transport and transit.
- Nuclear substances and radioactive sources possession, handling, and production licences.
- Nuclear and radiological emergencies.
- Nuclear and radiation safety.
- Nuclear safeguards and nuclear security.
- Civil liability for nuclear damage and penalties.

Law No. 7/2010 [II–2] is a comprehensive law covering safety, security and safeguards for regulatory control of all nuclear and radiological activities and facilities in Egypt.

II-3.2.2.   Challenges faced and lessons learned

The challenge was the preparation and approval of such a comprehensive nuclear law covering safety, security and safeguards for regulatory control of all nuclear and radiological activities and facilities in Egypt without conflict of responsibilities between ENRRA and other regulatory authorities (e.g. Ministry of Health, Environmental Affairs Agency) and without conflict with signed international legal treaties and conventions.

Amending the nuclear law enhanced the independence and effectiveness of ENRRA with clear responsibilities and regulatory functions related to nuclear and radiation safety and a distinction between the functions of ENRAA and the research functions.

The experience gained from operating and regulating Egyptian nuclear facilities and radiation facilities and activities supported the identification of all necessary elements of the national legal framework for the safety infrastructure including regulatory functions and responsibilities. This experience also helped in preparation and approval of the nuclear law (Law No. 7, 2010 [II–2]) and its amendment.

II-3.3. Regulatory infrastructure for safety

II-3.3.1. Implementation

Regulatory framework

The government recognized earlier the need for an effectively independent and competent regulatory body and considered the appropriate position of the regulatory body in the legal framework for safety.

The evolution of the Egyptian regulatory body [II–1]:

- 1982: Establishment of a competent regulatory body in Egypt (Nuclear Regulatory and Safety Committee).

- 1991: Presidential Decree No. 47, the National Centre for Nuclear Safety and Radiation Control (NCNSRC) was established as a part of the Egyptian Atomic Energy Authority (EAEA). (1992-1997) License/ permits for ETRR-2 research reactor were issued by National Centre for Nuclear Safety and Radiation Control (NCNSRC):

  - Construction Permit No. CP-01 for ETRR-2 reactor (basis Doc.: Safety Evaluation Report for ETRR-2 related to the construction permit);

  - Fuel loading permit for ETRR-2 reactor (November 1997);

  - Permissions to raise the reactor power up to 22 MW (1998);

  - Preliminary operation licence for ETRR-2 (September 1998) based on review and assessment of final safety analysis report of the reactor and results from the commissioning programme, and letter requesting operation licence, including information and documents to support the issuance of the licence.

- 2010 issuance of Law No. 7/ 2010 [II–2], an independent "Nuclear and Radiological Regulatory Authority (ENRRA)" was established.

- 2011: The executive regulation for the Law for regulating Nuclear and Radiological Activities promulgated by Law No. 7 of 2010 were issued according to the Prime Minister's Decree No. 1326 of 2011 [II–7]. The Executive Regulation is divided into nine chapters with 104 articles specifying the main rules, regulations and procedures governing the role of the ENRRA, as well the main responsibilities of the Chair of the Board and the responsibilities of ministries and agencies to ensure safety of transport of radioactive materials in the international context.

- 2012: ENRRA was established as a strong independent, transparent new entity to ensure the safe use of nuclear energy and ionizing radiation. The legal and regulatory framework, governing nuclear safeguards, safety and security of nuclear and radiation facilities, activities and practices in Egypt, includes a number of regulatory levels:

  - The nuclear law.
  - Executive regulation and governmental decrees.
  - Regulatory documents.
  - Industrial codes and standards.

The national regulatory body established in Egypt according to the Nuclear Law is ENRRA which carries out all regulatory and control works related to nuclear and radiological activities in Egypt for peaceful purposes of atomic energy, to ensure the security and safety of humans, property, and the environment from the harmful effects of ionizing radiation. The organizational structure of ENRRA was approved, whereby the prospective senior managers of the regulatory body are identified to meet the mandates of ENRRA as stipulated in Article 12 in the Nuclear Law [II–2].

Article 12 of the Nuclear Law stipulates ENRRA assigned responsibilities:
- Issuing and developing regulatory requirements and rules.
- Issuing, modifying, suspending, renewing, withdrawing and cancelling all types of licenses and permits.
- Reviewing and assessing the licensing safety related documents.
- Conducting regulatory inspection of all nuclear and radiological activities at all stages.
- Carrying out safeguards' inspections for nuclear materials.
- Control of transportation of radioactive materials.
- Coordinating emergency response at national level.
- Coordinating with governmental and non-governmental bodies.
- Issuing quarterly and annual reports to the Public

ENRRA is a governmental organization funded from the government and has income from licensing of nuclear and radiation facilities, activities and personnel. ENRRA provides quarterly and annual reports on its activities to the office of the Prime Minister and relevant ministries. A version of the annual report is also provided to the general public.

Ensuring resources of funding radioactive waste management and spent fuel management, the decommissioning of NPPs and the disposal of radioactive waste are considered in the Nuclear Law and its executive regulation [II–2].

II-3.3.2. *Challenges faced and lessons learned*

Experience gained through the regulating of nuclear facilities of Egypt including the research reactors and fuel manufacturing plant and radiological activities helped in identifying all necessary elements of the regulatory framework including NPP.

II-3.4. Other governmental considerations – global nuclear safety regime

II-3.4.1. *Implementation*

Experience gained through the construction, operation and regulation of Egypt's research reactors is helping in fully understanding the long-term commitments needed for an NPP, including accessibility to appropriate international legal treaties and conventions.

Egypt is party to the following international legal instruments adopted under the IAEA auspices (INIR report 2019) and preparing for the ratification of the CNS during 2023 [II–1]:

- Convention on Early Notification of a Nuclear Accident [II–9];
- Convention on Assistance in the Case of a Nuclear Accident or Radiological Emergency [II–10];
- Vienna Convention on Civil Liability for Nuclear Damage [II–11];

- Joint Protocol relating to the application of the Vienna Convention and the Paris Convention [II–12].

II-3.4.2. Challenges faced and lessons learned

Experience gained through the construction, operation and regulation of Egypt's research reactors and other non-nuclear activities is helping in fully understanding the long-term commitments needed for an NPP, including accessibility to appropriate international legal treaties and conventions.

II-4. KEY PRIORITIES FOR THE GOVERNMENT AND THE REGULATORY BODY IN EGYPT IN PHASE 1

II-4.1. Site survey

II-4.1.1. Implementation

Licensing is an ongoing process, starting at the stages of sitting and site evaluation and continuing up to and including decommissioning and the release from regulatory control.

The licensing process for nuclear facilities including NPPs was developed based on the experience of ETRR-2 licensing. According to the executive regulation requirements [II–7], the regulatory decision on the site approval permit is based on the assessment results of the reports submitted by the applicant, including the site evaluation report (SER) of a nuclear installation and the environmental impact assessment (EIA) report, which was approved by the Environmental Affairs Agency (EEAA).

The potential sites in Egypt for NPP construction are identified and candidate sites are selected based on a set of defined criteria, at a regional scale and with the use of available data.

According to Article 13 of the Executive Regulation of the Nuclear Law, an application for approving the site shall be submitted to ENRRA, including the purpose of construction NPP enclosed with [II–7][II–7]:

- Data on NPP (or other nuclear facilities);
- Identification of the legal rights of the applicant;
- Site Evaluation Report
- Approved EIA report.

ENRRA established the national requirements and provided criteria for ensuring safety in site evaluation for nuclear installations, namely: Site Evaluation Requirements for Nuclear Installation, 2016 [II–13].

ENRRA did not have enough experience to develop review procedures and method for NPP site related documentation and to assess and decide on the site approval permit. So, the need was raised for the cooperation with IAEA and/or TSOs to conclude on El Dabaa NPP site permit. Egypt hosted an IAEA SEED mission to provide a review in support of ENRRA's assessment and EEAA approval of EIA report [II-4].

The site approval of El Dabba NPP was issued in March 2019 based on national expertise review and cooperation with IAEA.

II-4.1.2.   Challenges faced and lessons learned

ENRRA did not have enough experience to develop review procedures and method for NPP site related licence documentation. The IAEA SEED mission provided a review in support of ENRRA's assessment. EEAA approval of EIA report was one of the conditions for ENRRA to grant the site approval permit which was issued in March 2019.

There were needs for cooperation with IAEA and/or external TSOs as an independent review to ensure that the staff were successfully conducting the SER review based on all relevant requirements of IAEA Safety Standards Series No. SSR-1 Site Evaluation for Nuclear Installations [II–14] , and before taking the regulatory decision on the site approval permit. The experience previously gained in review and assessment of the site and the coordination with EEAA, supported ENRRA's regulatory decision to and issuance of e the site approval permit for Spent Fuel Facility of El Dabaa NPP in 2022.

II-4.2. Consultation with interested parties

II-4.2.1.   Implementation

The Government of Egypt has always been committed to ensuring transparent, safe, and secure use of nuclear and radiological technologies; accordingly, a decision was taken in 2010 to create an effective legal framework that reflects Egypt's commitment towards its national and international obligations.

As per the Egyptian Nuclear Law on Regulation of Nuclear and Radiological Activities (Law No. 7/2010 [II–2]), its amendments, and the Executive Regulations of the law, ENRRA is responsible for informing the public about the regulatory process of nuclear and radiation activities, setting means and procedures to involve the public, ensuring necessary procedures are taken to disseminate nuclear safety and security culture, and providing the public with any necessary information about the status of nuclear and radiation safety in the area of their residence unless this information is confidential by nature [II.2].

ENRRA's communication team has evaluated ENRRA's communication strategy and plan for year 2022 and analysed clearly points of strength, weakness, what was achieved and what still needs to be achieved. Based on this evaluation and lessons learnt ENRRA's communication strategy and plan for year 2023 were drafted.

Interested parties' mapping

Many relevant interested parties are involved in the decision-making process to gain public confidence in the nuclear regulatory system in Egypt. ENRRA, as the only regulatory body in Egypt, has a wide array of interested parties with different needs about the information required, including:

- Public;
- Non-governmental organizations;
- Operators and licensees;
- International organizations;
- Government;
- Media;

- Medicals and professionals;
- Manufactures and contractors;
- Regulatory bodies;
- Research institutes and universities;

Key communication tools

The ENRRA team is aware of different and various communications tools and techniques. Identifying and prioritizing tools that are used in different phases and situations will depend on three main factors: Resources (financial, human, communications tools), speed of delivery, and control of messages that will be delivered.

The most important tools are:

- Participating in person in outreach activities and meetings with representatives of interested parties;
- Site visits;
- Public meetings and hearings;
- Workshops;
- Interviews;
- Digital presence through various tools;
- Website;
- Emails;
- Social media platforms;
- Internet;
- Public consultations;
- Bilateral agreements;
- Memorandums of Understanding;
- Printed materials (brochures, newsletters, articles in newspapers and magazines)
- Publications;
- Billboards, banners, news and television advertisements;
- Press releases;
- Videos.

ENRRA has a strategic goal to continuously improve the engagement of interested parties and public awareness. It has organized workshops to raise awareness about nuclear power and the role of ENRRA for the public and media.

ENRRA's website provides information regarding the organization's vision and mission, and general information regarding licences. ENRRA provides quarterly and annual reports on its activities to the office of the Prime Minister and relevant ministries. A version of the annual report is also provided to the public [II–1]:.

II-4.2.2.   Challenges faced and lessons learned

Based on an annual evaluation of the communication strategy and plan:

- Public interaction through social media tools was not sufficient, new tools — in addition to other tools —were introduced since the public is the core of any strong, transparent regulation of the nuclear industry. Establishing and maintaining public confidence is a priority.

- Communicating with Government entities and officials; using reporting and MOUs showed effectiveness and efficiency in communicating with government officials.

II-4.3. Leadership and management for safety

II-4.3.1. *Implementation*

ENRRA in relation to transitioning to Phase 2, has drafted its Management System Manual. It includes ENRRA's vision, mission and policy, a description of the structure of the management system, a description of the main processes, the organizational structure and functions, the importance of leadership and safety culture, allocation of resources and training, and department objectives [II-1].

ENRRA issued the Management System Requirements for Regulated Facilities and Activities in 2016 [II–15] based on IAEA Safety Standards Series No. GSR Part 2, Leadership and Management for Safety [II–16]. ENRRA management participated in several events on safety and security culture and has invited relevant interested parties, including NPPA.

ENRRA has established an integrated management system (IMS) including safety culture policy and safety culture plan. ENRRA's process map contains the following core processes:

- Development of regulations and guides;
- Authorization;
- Regulatory inspections of facilities and activities;
- Review and assessment of facilities and activities;
- Enforcement;
- Communication and consultation with interested parties;
- Emergency preparedness, coordination and response.

How safety culture was fostered

The ENRRA management gives priority to the development of safety culture at all levels of ENRRA organization. The activities of ENRRA management are aimed to creating an atmosphere of understanding and commitment of all ENRRA personnel to safety, as follows:

- ENRRA conducts a self-assessment of the safety culture level and annually draws up an action plan to develop ENRRA personnel safety culture.
- The procedure for creation, maintenance and development of ENRRA safety culture and self-assessment of safety culture was described in more detail.
- ENRRA appointed a safety culture officer, responsible for organizing safety culture self-assessment for personnel, and for monitoring the implementation of the action plan for the safety culture development.
- The ENRRA management supported and appreciated any personnel initiatives aimed at enhancing safety.

Many challenges exist to ENRRA safety culture which are recognized, understood and overcame.

II-4.3.2. *Challenges faced and lessons learned*

Coordination with all other organizations involved in the nuclear emergency management is a challenge. The operation and regulation of Egypt's Second Research Reactor ETRR-2) and other nuclear and radiation facilities support identifying response organizations and arrangements for emergency preparedness and response. ENRRA headed a Supreme Committee for Nuclear and Radiological Emergencies ready for coordination with all other organizations involved in responding to an emergency with sufficient means and the appropriate competence and leadership.

The challenges to ENRRA safety culture include:

- Maintaining the focus on safety under external pressure from interested parties;
- Considering economic factors and budget limitations, and keeping appropriate resources;
- Interfacing with other regulations and regulators.

ENRRA recognized, understood and overcame these challenges through:

- Demonstration of leadership for safety at all levels in ENRRA. All leaders prioritize the consideration of safety over other matters and demonstrate a commitment to safety in their decisions and behaviours. Leaders clearly define individual roles, responsibilities and authority, and ensure the availability of sufficient resources.
- Open and transparent communication with all interested parties building trust and confidence, including information concerning regulatory activities, and the decision-making process.
- Clear regulatory framework without undue constraints. Regulatory bodies focus on and encourage the licensee to achieve higher levels of safety and have the primary responsibility for safety. The regulatory framework takes into consideration interfaces with other regulations and regulatory bodies.
- Continuous improvement, learning and self-assessment at all levels in the organization. Continuous improvement and learning including arrangements for formal training, implementing best practices, and participating in international activities. Self-assessment of the safety culture supports continuous improvement.

This helped in turning these challenges into opportunities to further strengthen the safety culture of ENRRA.

II-4.4. Development of human resources

II-4.4.1.   Implementation

Human resources are under development in areas relating to nuclear safety by participation of Egyptian engineers and scientists in: reactor design (general design and specific reactor); components manufacture and performance tests and qualifications by prototypes, mock-ups and tests; construction, commissioning and initial (pre-operational) tests; and operation and maintenance and modification.

Research reactors prepare and qualify the Egyptian cadres for the management of nuclear power plants, including planning, design, construction, commissioning and licensing and operation, and licensing and regulatory oversight of nuclear installation. ENRRA staff involved in

- Licensing of research reactor construction, commissioning, and operation
- Licensing and re-licensing of ETRR-2 operators
- Conducting inspection according to regulatory inspections programme for Egyptian research reactors

The licensing and conducted inspections on Egypt research reactors provided training and re-training for a number of experts involved in ensuring the safety of nuclear installations and later in identifying the required competences and number of experts and identifying gaps in safety related training. This ensure that senior managers and prospective safety experts to be involved in the nuclear power programme gain an understanding of the principles and criteria of nuclear safety.

ENRRA developed human capabilities in areas of nuclear installation safety by the following means:

- Senior managers and specialists have experience and are engaged in the ENRRA activities.
- A number of senior managers and specialists come from nuclear facilities.
- Second step is to complete the organization with freshly graduated officers.

External support organizations and contractors

The availability of expertise that could support the safety infrastructure in the long term are considered later in contracting an international support organization. Building on experience of regulating nuclear facilities and international and vendor practices, areas of cooperation and detailed scope of services are negotiated and identified. Areas of technical support and scope of services are:

- Development of the regulatory framework;
- Development of an integrated management system (IMS) and electronic management system (EMS);
- Human resources development and capacity building;
- Nuclear power plant authorization process;
- Planning and implementation of ENRRA regulatory inspection and SSC conformity assessment for nuclear power plants;
- Development of tools for effective coordination of emergency situations under the Situation and Analytical Centre (SAC).

Research for safety and regulatory purposes

As Egypt has research and development organizations dealing with safety research, a considerable level of expertise and in-depth knowledge had been acquired to assess and analyse safety related aspects of a nuclear power plant.

ENRRA identified research centres and universities for cooperation in research and analyses of safety related areas. The analyses are carried out to address issues and questions identified by the regulators as well as to benchmarks results.

II-4.4.2.   Challenges faced and lessons learned

ENRRA was established in early 2012 according to the Nuclear Law (Law No. 7/2010 [II–2]), as a separate entity from the research and development organization of nuclear and radiological safety. Initially, ENRRA employees performed two tasks: regulatory function and research and development. The challenge is to discern between the two tasks and enforce the effectiveness and independence of a regulatory body with clear responsibilities and relevant regulatory functions and responsibilities related to nuclear and radiation safety.

Currently, ENRRA's new organizational structure takes into consideration IAEA recommendation and lesson learned from other countries, including:

- Achieving full independence and autonomy without being isolated from the interested parties.
- Setting clear organizational structure with well-defined tasks and responsibilities.
- Recruiting, developing and retaining professionals.
- Establishing a competence management system.
- Adopting values and principles to guide ENRRA's mission toward safety openness, efficiency, clarity and reliability.
- Establishing an integrated management system that guides all procedures and processes of ENRRA.
- Establishing a knowledge management system to support the reservation and localization of the knowledge gained and support the sharing of ENRRA practices and experiences with other countries.

REFERENCES TO ANNEX II

[II–1]   INTERNATIONAL ATOMIC ENERGY AGENCY, MISSION REPORT ON THE INTEGRATED NUCLEAR INFRASTRUCTURE REVIEW (INIR) - PHASE 2 (2019). https://www.iaea.org/sites/default/files/documents/review-missions/inir2-egypt.pdf

[II–2]   THE ARAB REPUBLIC of EGYPT, Law No. 7 of the Year 2010 On the Enactment of the Law on Regulation of Nuclear and Radiation Activities, Official Journal, Issue No. 12 (Bis-A), Cairo (2010). https://enrra.org/publication/law-regulating-nuclear-radiological-activities/

[II–3]   EGYPTIAN NUCLEAR AND RADIOLOGICAL REGULATORY AUTHORITY (ENRRA), Regulatory Procedures on Inspection of Nuclear installations in Operating Stage, Cairo (2016). https://enrra.org/publication/classification/

[II–4]   INTERNATIONAL ATOMIC ENERGY AGENCY, IAEA team of experts has concluded an eight-day Site and External Events Design (SEED) review mission to Egypt, IAEA, Cairo (2019). https://www.iaea.org/newscenter/pressreleases/an-international-atomic-energy-agency-iaea-team-of-experts-has-concluded-an-eight-day-site-and-external-events-design-seed-review-mission-to-egypt

[II–5]   ARAB REPUBLIC OF EGYPT, Technical Assistance to Support the Reform of the Energy Sector, Executive Summary, Integrated Sustainable Energy Strategy to210, Volume 1, November 2015, Cairo (2015).

[II–6]   THE ARAB REPUBLIC of EGYPT, Law No. 210 of 2017 Amending Some Provisions of Law No. 13 of 1976 on the Establishment of the Nuclear Power Plants

Authority for Generating Electricity, Presidency of the Arab Republic of Egypt, 29 November 2017, Cairo (2017).

[II–7] Prime Ministerial Decree No. 1326 of the Year 2011 On the Issuance of the Executive Regulations Implementing the Law on Regulation of Nuclear Activities and Radioactivity Promulgated by Law No. 7 of the year 2010, Official Journal, Issue No. 42 (Bis), 26 October 2011, Cairo (2011).

[II–8] INTERNATIONAL ATOMIC ENERGY AGENCY, Country Nuclear Power Profiles-Egypt (2021). https://www-pub.iaea.org/MTCD/Publications/PDF/CNPP-2021/countryprofiles/Egypt/Egypt.htm

[II–9] Convention on Early Notification of a Nuclear Accident, INFCIRC/335, IAEA, Vienna (1986).

[II–10] Convention on Assistance in the Case of a Nuclear Accident or Radiological Emergency, INFCIRC/336, IAEA, Vienna (1986).

[II–11] Vienna Convention on Civil Liability for Nuclear Damage, INFCIRC/500, IAEA, Vienna (1963).

[II–12] Joint Protocol Relating to the Application of the Vienna Convention and the Paris Convention INFCIRC/402, IAEA, Vienna (1988).

[II–13] EGYPTIAN NUCLEAR AND RADIOLOGICAL REGULATORY AUTHORITY (ENRRA), Regulatory Requirement, Site Evaluation Requirements for Nuclear Installation, Document ID: ENRRA-NF/RR/SP-00, Rev. No. 00, 29 February 2016, Cairo (2016). https://enrra.org/publication/site-evaluation-requirements-for-nuclear-installations/

[II–14] INTERNATIONAL ATOMIC ENERGY AGENCY, Site Evaluation for Nuclear Installations, IAEA Safety Standards Series No. SSR-1, IAEA, Vienna (2019).

[II–15] EGYPTIAN NUCLEAR AND RADIOLOGICAL REGULATORY AUTHORITY (ENRRA), Management System Requirements for Regulated Facilities and Activities, Document ID: ENRRA-GE/RR/MG-00, Rev. No. 00, 8 March 2016, Cairo (2016). https://enrra.org/publication/facilities-management-system/https://enrra.org/publication/facilities-management-system/

[II–16] INTERNATIONAL ATOMIC ENERGY AGENCY, Leadership and Management for Safety, IAEA Safety Standards Series No. GSR Part 2, IAEA, Vienna (2016), https://doi.org/10.61092/iaea.cq1k-j5z3

ANNEX III.
CASE STUDY ON GHANA

III-1. EXISTING SAFETY INFRASTRUCTURE PRIOR TO INITIATING A NUCLEAR POWER PROGRAMME IN GHANA

The effort to have nuclear power in Ghana dates back to the first President who acknowledged the vast potential of nuclear technology to provide reliable source of electricity for the countries of the world in 1963. The deployment of the Miniature Neutron Source Reactor (MNSR) through assistance received from the International Atomic Energy Agency and a trilateral agreement with China Institute of Atomic Energy (CIAE) in November 1994 led to the establishment of the Radiation Protection Board (RPB). The Atomic Energy Act 204 of 1964 established the Ghana Atomic Energy Commission (GAEC) [III-1]. The National Nuclear Research Institute (NNRI) was the operating organization of Ghana Research Reactor 1 (GHARR-1) and the RPB was the regulatory body that provided regulatory oversight for the operation of the reactor. The LI 1559 has since been repealed by the Nuclear Regulatory Authority Act, 2015 (Act 895) [III-1]. The operational functions of the RPB were carried out by the Radiation Protection Institute of GAEC, which was established to provide technical support for the enforcement of the legislative instrument. The RPB developed guidelines GRPB-G1 on Qualification and Certification of the Radiation Protection Personnel, GRPB-G2 on Notification and Authorization by Registration or Licensing [III-1], Exemption and Exclusions, GRPB-G3 on Dose Limits and GRPB-G4 on Inspection to provide practical guidance in respect of LI1559.

Both the NNRI and RPB were provided annual budgetary allocations by the Government of Ghana for the operation and regulation of the reactor. To ensure the safety of GHARR-1, the RPB performed a variety of activities, including the development and documentation of the licensing bases that specified design requirements and operation and maintenance practices; the inspection and enforcement of licence requirements; the performance of technical research and analysis; and the modification of regulatory requirements as needed. All these activities were geared towards the safe operation and regulation of GHARR-1 and other facilities and activities involving radiation in Ghana.

Although the RPB played a central role in assuring safety of the reactor, the primary responsibility for the safe operation of GHARR-1 rested with the NNRI, the licensee. The NNRI was ultimately responsible for the design, operation, and maintenance of the reactor — not only to meet RPB requirements but also to assure safety. To pool resources, share experiences, and coordinate efforts, the NNRI collaborated with the IAEA to provide technical assistance for the safe operation of GHARR-1. The RPB also conducted a series of inspections to ascertain the level of adherence to the requirements.

At the operating organization level, responsibilities that accompanied the development and implementation of the quality assurance programme were anchored in the management of GHARR-1 providing the means and support to achieve objectives; personnel performing work to achieve quality; and evaluation of the effectiveness of management processes and work performance. The Radiation Safety Committee (RadSC) and the Reactor Safety Committee (RSC) of GAEC provided safety advisory support for operation of the GHARR-1. Experiments involving a transfer of irradiated samples outside the reactor facility were reviewed by RSC and RadSC and approved by the RPB. The Radioactive Waste Management Centre provides support to the reactor facility in storing solid waste in accordance with requirements. In 2008, the Government of Ghana took a cabinet decision to add nuclear power to the energy mix of the country. In 2010, nuclear power was included in the National Energy Policy. A nuclear power unit was then established at the Ministry of Energy and Petroleum in 2012, and the Ghana Nuclear Power Programme Organization (Ghana's NEPIO) was inaugurated the same year (2012) chaired by the Deputy Minister of Energy in charge of Power. Eight technical groups comprising Techno-Economic Assessment and Financing Group, Nuclear Power Plant Technology Assessment Group, Siting and Grid Infrastructure Assessment Group, Legal Group, Regulatory Group, Human Resource Development Group, Environmental Assessment Group and Nuclear Power Project Management and Stakeholders' Involvement Group were formed to support activities of the Ghana Nuclear Power Programme Organization (GNPPO). Several activities, in connection with the IAEA Milestones approach [III-3] were carried out including public education on the nuclear option through radio and television programmes and newspapers; development of licensing procedures by the RPB; development of NPP technology assessment reports; creation of a Nuclear Power Centre at GAEC to coordinate and harmonize the activities of the eight technical groups; discussion and signing of MOUs and agreements between Ghana and interested countries; zoning of suitable areas for nuclear power plant siting; and drafting and promulgation of the nuclear law.

Transition from research reactor to NPP

The infrastructure developed in the preparatory works including planning, negotiation, and training set the basis for initiating the steps for developing infrastructure for nuclear power in Ghana.

The construction and commissioning of the miniature neutron source reactor with assistance from the CIAE provided hands-on experience for staff of the operating organization and the RPB. The training accorded to the staff at CIAE facilities and in Ghana provided the basis for authorizing the research reactor.

The maintenance oversight of the GHARR-1 facility has provided opportunities for training staff of NRA on its possible use in the nuclear power programme.

III-2. STATUS OF NUCLEAR POWER PROGRAMME IN GHANA

Ghana has enacted the Nuclear Regulatory Authority Act, 2015 (Act 895) [III-4], allowing for formation of Nuclear Regulatory Authority (NRA) addressing regulations and management of activities and practices for the peaceful use of nuclear energy and radiation, management of radioactive waste and spent fuel resulting from civilian applications, management of radioactive waste and spent fuel resulting from civilian applications and liability for nuclear damage. Ghana has hosted a Phase 1 Integrated Nuclear Infrastructure Review (INIR) mission in January 2017 [III-1] and its follow-up mission in October 2019 [III-2]. The Programme Comprehensive Report (PCR) was submitted to enable the Government to take a decision on the nuclear power programme and an Inter-Ministerial Committee was constituted to address funding options and interacting with neighbouring countries on introduction of nuclear power in Ghana. The Government announced the decision to embark on nuclear power in August 2022. Large reactors and recently small modular reactors are under consideration by the Nuclear Power Ghana (owner/operator). There is one research reactor, GHARR-1, which started operation in March 1995. A five-year plan is being implemented to equip the NRA to effectively conduct oversight activities of the nuclear power programme. Draft regulations to ensure nuclear safety, security and safeguards are being developed. The NRA is developing an integrated management system in line with IAEA Safety Standards Series No. GSR Part 2, Leadership and Management for Safety [III–5] and ISO 9001:2015 [III–6]. The NRA is implementing a European Instrument for International Nuclear Safety Cooperation (INSC) project for nuclear regulatory infrastructure development. Bilateral arrangements have been made with the United States Nuclear Regulatory Commission, Canadian Nuclear Safety Commission and Pakistan Nuclear Regulatory Authority and support is being received from the International Atomic Energy Agency and United States Department of Energy.

III-3. ROLES AND RESPONSIBILITIES OF THE GOVERNMENT AND THE REGULATORY BODY IN GHANA IN PHASE 1 OF THE NUCLEAR POWER PROGRAMME

The Government of Ghana formed the GNPPO which comprises all organizations with roles in the programme such as GAEC, the regulatory body, the Ministries of Energy, the Environment, Science, Technology and Innovation, Interior, Defence and National Security, the Volta River Authority, the Bui Power Authority, the Geological Survey Authority, the Ghana Meteorological Agency, the Environmental Protection Agency and the Water Resources Commission. The Government enacted the Nuclear Regulatory Authority Act, 2015 (Act 895) and formed the NRA. The Government has also formed Nuclear Power Ghana as the owner/operator.

III-3.1. National policy and strategy for safety

III-3.1.1. Implementation

The National Policy for Nuclear and Radiation Safety is currently in draft form in Ghana. It draws inputs from the Nuclear Regulatory Authority Act, 2015 (Act 895) [III-1] and addresses the requirements of Ghana under the international nuclear safety conventions and relevant requirements and guidance of the IAEA. It was developed based on GSR Part 2 [III–5]], and IAEA Safety Standards Series Nos GSG-12, Organization, Management and Staffing of the Regulatory Body for Safety [III–7] and GS-G-3.1, Application of the Management System for Facilities and Activities [III–8]. It sets out expectations for behavioural requirements for leadership, responsibilities for development and implementation of a management system, commitment to planning, strategy development and implementation by parties, monitoring, evaluation and improvement, human and organizational factors. It encourages a questioning attitude of staff, and provides a reflection of safety, security and safeguards attributes, graded approach, and integration of the management system.
The Policy calls for active involvement in research and development to promote nuclear and radiation safety. The Policy will serve as basis for development of organizational goals, strategies, plans and objectives in relation to nuclear and radiation safety. The Policy addresses openness, transparency, and mutual trust, subject to legal requirements on provision of information, protection of sensitive information and consultation with interested parties. The Policy addresses safety, health, environmental, security, societal, quality and economic factors. The Policy aligns with and support the development of a strong safety culture.

The strategic plan of the NRA is purposed to resource the Authority with strategic direction that will facilitate its operation. The plan translates the vision and mission of the NRA, drawn from its mandate, statutory functions and the focus of the management, into strategies and priority actions to be implemented over a period of five years to enable the NRA to fulfil its statutory obligations. The plan focuses on key objectives, strategies and actions to be implemented in the medium term to ensure that humans and the environment are protected from all harmful effects of radiation. The plan serves as basis to track the growth of the NRA over the short, medium and long term. To implement the strategic, plan two categories of strategy elements have been adopted: general strategy elements and basic and specific functions. The general elements identified are stable legislative framework; functional separation; management system; competent staff; budget; prime responsibility for safety; graded approach; technical support organizations; record keeping; and information system. The basic and specific functions are licensing; review and assessment; inspection and enforcement; development of regulatory tools; emergency preparedness and response; management of radioactive waste, spent fuel and decommissioning; interfaces of safety, security and safeguards; public information and outreach; international cooperation; transparency, openness and trustworthiness; nuclear security detection architecture.

The regulatory strategy was developed with support from the European Instrument for International Safety Cooperation (INSC) Project. A questionnaire based on IAEA Safety Standards Series Nos GSR Part 1 (Rev. 1), Governmental, Legal and Regulatory Framework for Safety [III–9] and SSG-16 (Rev. 1), Establishing the Safety Infrastructure for a Nuclear Power Programme [III–10] were used to describe the status of the governmental, legal and regulatory framework for safety following which a series of virtual workshops and on-site-assistance were held to discuss gaps, European experiences and performance indicators and regulatory strategic approaches.

III-3.1.2. Challenges faced and lessons learned

The National Policy and Strategy for Nuclear and Radiation Safety was developed based on the initial draft of the National Radiation Safety Policy. It focuses attention on safety and security culture demonstration in the conduct of regulatory oversight of the use of nuclear and other radioactive materials and activities in Ghana. It supports implementation of the provisions of the Nuclear Law. The NRA has a team working on a Nuclear Safety and Security Culture Programme.

III-3.2. Governmental and legal framework for safety

III-3.2.1. Implementation

Nuclear law

The GAEC Legal Office led the drafting of the Nuclear Law in Ghana [III-2]. The Law was drafted with the support of the IAEA Office of Legal Affairs through scientific visits and fellowships for Ghanaian legal experts. With the focus of the RPB Legislative Instrument on radiation safety and radiation protection, the assessment focused on domesticating international legal instruments relevant for nuclear power in areas of nuclear safety, nuclear security, safeguards, non-proliferation and liability for nuclear damage. The drafting of the nuclear law coincided with adoption of the relevant international legal instruments. To enable nuclear law integration in the nuclear education in Ghana, students at the School of Nuclear and Allied Sciences were introduced to nuclear law and legislation through an intensive course organized with support of the Agency. The course on nuclear law and legislation has since been included in the curriculum of the school for all students. The Law is under review to address pertinent areas identified following its implementation since January 2016.

III-3.2.2.   Challenges faced and lessons learned

The existence of the RPB Legislative Instrument facilitated the development of the comprehensive nuclear law, which is currently under review.

III-3.3. Regulatory infrastructure for safety

III-3.3.1.   Implementation

Regulatory framework

The need for effectively independent and competent regulator was carefully considered in the appointment of staff. Staff of the Radiation Protection Institute (RPI) and National Nuclear Research Institute (NNRI) of GAEC were transferred to form the NRA. The scientific staff comprised two Chief Research Scientists, one Principal Research Scientist, one Senior Research Scientist, fourteen Research Scientists, twenty Assistant Research Scientists, one Chief Technologist, one Principal Technologist, two Technologist and one Technician. The administrative staff comprised one Senior Auditor, one Legal Officer, one Administrative Officer, one Chief Administrative Assistant, one Principal Accounting Assistant, one Principal Administrative Assistant, two Senior Administrative Assistants, two Drivers, one Farm Supervisor and one Departmental Assistant. In appointing senior managers, the conditions at GAEC for similar ranks was carefully followed to select managers who could demonstrate leadership as well as those knowledgeable in the areas they were being assigned to initially act as senior managers of the NRA. Those to be transferred as senior managers were engaged individually to ascertain their readiness to join the new regulatory body. A memorandum of understanding was signed between GAEC and the NRA to formalize the transfer of staff to form the NRA. Equipment of RPI for regulatory purposes was transferred to the NRA. The NRA has continued to use the Conditions of Service and Scheme of Service of GAEC, while a new framework tailored to our specific needs is currently being developed. Staff continue to contribute to the Pension Scheme of GAEC and benefit from same. Radioactive Waste Management Centre of GAEC continues to serve the users of radiation across the country and supports NRA activities in relation to the handling of orphan sources.

The NRA has been placed under the Ministry of Environment, Science, Technology and Innovation (MESTI) to enable the Authority to join the likes of the Environmental Protection Agency and the Land Use and Spatial Planning Authority in the same Ministry. The Ministry of Energy is the main promoter of the nuclear power programme in Ghana, with active support from GAEC.

The Government obtains advice from the NRA and consults with the NRA in relation to nuclear and radiation safety of the nuclear power programme. In forming the owner/operator company for the nuclear power programme, the NRA was engaged through the GNPPO to be aware of the efforts and to enable the NRA to engage the owner/operator on its responsibilities for nuclear and radiation safety in relation to the programme.

The NRA has adopted a hybrid approach between prescriptive and performance approaches to regulatory infrastructure development. This optimizes the use of the limited number of staff and the roles to ensure effective use of experiences of other regulatory bodies in similar roles.

III-3.3.2. Challenges faced and lessons learned

The experience with research reactor operation and documentation handling, regulatory guidelines development, regulatory oversight, independent safety oversight supported the implementation of NRA roles in Phase 1 of the nuclear power programme. The NRA is preparing to review applications, regardless of the technology chosen.

III-3.4. Other governmental considerations – global nuclear safety regime

III-3.4.1. Implementation

International legal instruments

Ghana is a party to the international legal instruments needed for introducing nuclear power such as Convention on Nuclear Safety [III–11], Joint Convention on the Safety of Spent Fuel Management and on the Safety of Radioactive Waste Management [III–12], Convention on Assistance in the Case of a Nuclear Accident or Radiological Emergency[III–13], Convention on Early Notification of a Nuclear Accident [III–14], Convention on Supplementary Compensation for Nuclear Damage [III–15]. Efforts are underway to ratify the International Convention on Suppression on Acts of Nuclear Terrorism (ICSANT) and Treaty on the Prohibition of Nuclear Weapons [III–16]. Ghana has adopted the Code of Conduct on the Safety and Security of Radioactive Sources [III–17], the Code of Conduct on the Safety of Research Reactors [III–18] and the Guidance on the Import and Export of Radioactive Sources of the International Atomic Energy Agency[III–19].
Most provisions in the above instruments and codes have been included in the Nuclear Law. A review of the Law is currently underway to address the areas which are not comprehensively addressed [III-4].

Convention on Nuclear Safety

The NRA facilitated the preparation of Ghana's National Report to the Convention on Nuclear Safety in 2016 and coordinated the provision of responses to the questions raised by the other Contracting Parties on the report [III-4]. The Acting Director-General of the NRA, Prof. G. Emi-Reynolds served as one of the two Vice Presidents for the 7th Review Meeting held on 27 March to 7 April 2017. The NRA facilitated the review of reports of other Contracting Parties and led the discussion on Ghana's report at our Country Group.

Joint Convention

The NRA participated in the Sixth Review Meeting of the Contracting Parties of the Joint Convention on the Safety of Spent Fuel Management and on the Safety of Radioactive Waste Management held at IAEA Headquarters from 21 May to 1 June 2018. The NRA participated in the preparation of the National Report, the review of reports of other Contracting Parties, provision of responses to questions posed on Ghana's report and also in the discussion on Ghana's report at our Country Group [III–20].

Ghana is a founding member of the Forum of Nuclear Regulatory Bodies in Africa (FNRBA). The NRA currently serves as Secretary to the Forum and coordinates Thematic Working Group 2 on Radiation and Waste Safety of the Forum. The NRA previously coordinated the Thematic Working Group on Research Reactor Safety through which various activities were held to address nuclear safety needs of research reactors in Africa.

Ghana participates actively in the Global Nuclear Safety and Security Network (GNSSN) and participates in the GNSSN side events at the Agency General Conferences.

When the PCR was submitted to the Government, an inter-ministerial Committee was formed by the President comprising the Minister of Foreign Affairs and Regional Integration, Minister of Finance, and Minister of Environment, Science, Technology and Innovation to inform neighbouring countries of Ghana's intention to construct a nuclear power plant and provide strategy for funding the project prior to public announcement of the intention. The public announcement has since been made.

International Cooperation

Section 5(o) of Act 895 mandates NRA to exchange information and cooperate with regulatory authorities of other countries and relevant international organizations on matters of nuclear safety, nuclear security and safeguards [III-1]. The NRA is participating actively in the FNRBA, Regulatory Cooperation Forum (RCF) and IAEA regulatory activities. An Arrangement for Technical Information Exchange and Cooperation in Nuclear Safety Matters has been signed with the United States Nuclear Regulatory Commission (USNRC) following which an Agreement for Radiation Protection Computer Code Analysis and Maintenance Programme (RAMP) has also been signed. The NRA has working relations with the United States Department of Energy on nuclear security and safeguards. The NRA is implementing support under the INSC while staff benefited from training and tutoring activities of the European Nuclear Safety Tutoring and Training Institute (ENSTTI) and are currently benefiting from ongoing tutoring and training activities and European Leadership for Safety Education (ELSE) activities of the European Union. The INSC project is addressing regulatory strategy, and its practical implementation, for enhancing the capacity and effectiveness of the NRA; development of an integrated management system; human resources development plan and training programme; basic safety principles, requirements and criteria for new nuclear reactors; enhancing the capabilities of the NRA for site licensing activities; and transparency and public information. The NRA has signed a Memorandum of Understanding with the Canadian Nuclear Safety Commission (CNSC) and Pakistan Nuclear Regulatory Authority (PNRA). There is ongoing discussion with Rostekhnadzor of the Russian Federation (RTN) and Moroccan Agency for Nuclear and Radiological Safety and Security (AMSSNUR) of Morocco for collaboration.

III-3.4.2. *Challenges faced and lessons learned*

The NRA has benefited from the participation in the Convention on Nuclear Safety and Joint Conventions which have supported in highlighting the challenges and areas of good performance. These have been used to improve the regulatory infrastructure development in Ghana.

III-4. KEY PRIORITIES FOR THE GOVERNMENT AND THE REGULATORY BODY IN GHANA IN PHASE 1

III-4.1. Site survey

III-4.1.1. *Implementation*

The NRA is developing the Site Evaluation Regulations to enable the issuance of site licences. A team of eight has been assigned the role of leading the regulatory oversight of siting by the NRA. The siting oversight is being conducted as a project and a charter has been issued to the

Team which addresses requirements as addressed in Act 895, commitments under Convention on Nuclear Safety and the Vienna Declaration on Nuclear Safety, resources [III-4], list of key interested parties, interested party requirements, assumptions, constraints and risks. The deliverables are site evaluation regulations, guidelines, work breakdown structure, standard review plan, and review of safety, security and safeguards conditions around the site.

The NRA hosted USNRC- International Regulatory Development Partnership (IRDP) Workshop on Site Application Review for Nuclear Power Plants from 16 to 20 October 2017. The NRA participated in the Site and External Events Design (SEED) expert mission of the IAEA held from 27 to 30 November 2017. The NRA hosted an IAEA National Workshop on Review of Site Licence Submittals in Ghana Nuclear Programme from 23 to 27 January 2023. A Guideline has been issued to Nuclear Power Ghana on the content of the Site Approval Report, which addresses site selection criteria process, geography, demography, meteorology, hydrology, geology, seismology, geotechnical needs, radiological conditions, emergency planning, environmental, other non-safety related considerations, nuclear security and the management system. The NRA engaged the Nuclear Power Ghana (NPG) to provide inputs into the document, following which it was finalized. The siting oversight milestones and siting oversight breakdown structure are presented in Figs [III.1] and [III.2].

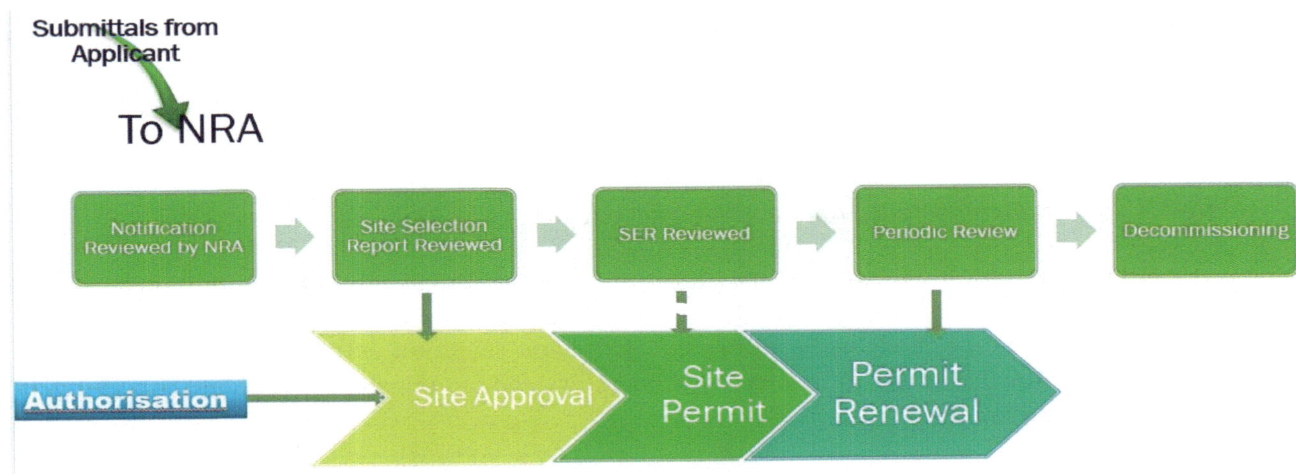

Fig. III.1: NRA siting oversight milestones

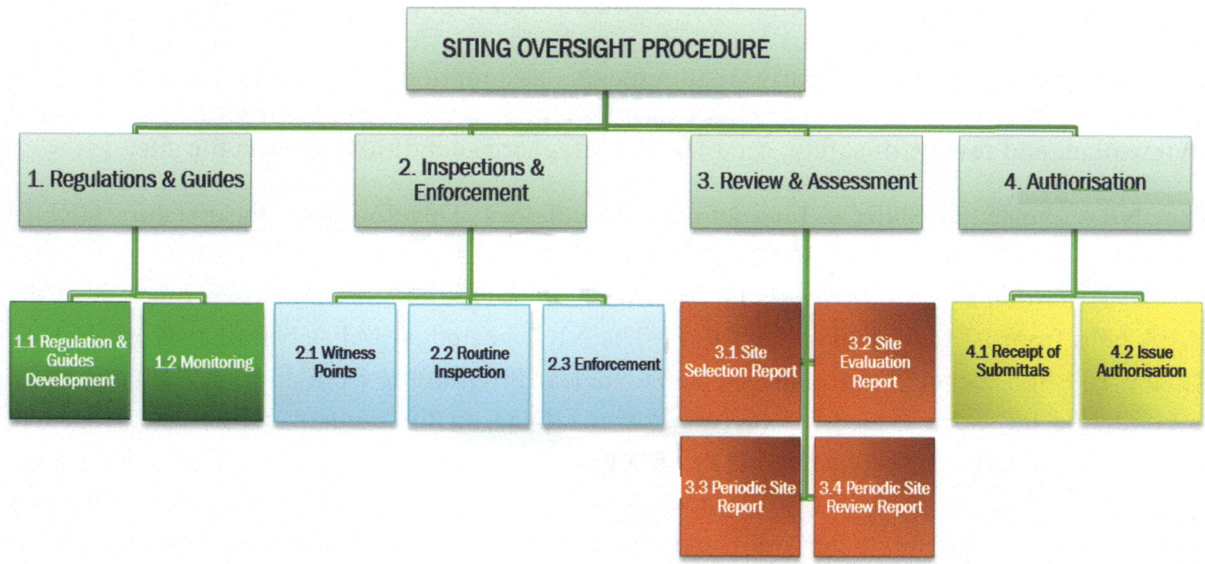

Fig. III.2: Siting oversight breakdown structure

Through the INSC Project, the site evaluation regulations were reviewed. A Site Licensing Activities Workshop was held virtually from 14 to 21 September 2020. A training course on seismic hazard applications in site evaluation was held from 6 to 10 June 2022. Two members of the siting team were taken through on-the-job training on siting submittals review in Hungary and Slovakia from 5 September to 28 October 2022.

III-4.1.2. Challenges faced and lessons learned

The NRA is developing the regulatory framework to enable licensing of sites for nuclear power plants in Ghana. The two-step approach is being implemented as required by the nuclear law. The regulatory body needs to be informed of the site selection activities to enable effective preparedness for the review of submittals. Operators need to engage the regulatory body early in the process to avoid delays.

III-4.2. Consultation with interested parties

III-4.2.1. Implementation

Transparency and openness

The NRA supports efforts of the GNPPO in informing the public and other interested parties of benefits and risks of nuclear power. The NRA on its own continues to engage the public in nuclear and radiation matters. Trainings have been organized for media personnel with specialty in scientific reporting as well as media outlets with wide coverage and reach across the country. The NRA focuses on discussing the role of the nuclear regulatory body in the nuclear power programme and development of regulatory infrastructure in Ghana. Some of the interested parties of the NRA are the Environmental Protection Agency, the National Disaster Management Organization, the Ghana Atomic Energy Commission and the Ghana Standards Authority. The NRA involves staff of these organizations in the training activities obtained through the IAEA and other international partners.

Public engagement activities

Section 5(k) of the Nuclear Regulatory Authority, 2015 (Act 895) [III-4], requires the NRA to educate the public on nuclear and radiation matters. The areas of concern are branding, visibility and building the corporate image of NRA, educating the public on nuclear and radiation matters, regulatory communications research (Act 895, 5(d)), stakeholder engagement (Act 895, 5 (h)), and regulatory and emergency communications (Act 895, 5 (i)).

The NRA is creating awareness on the establishment of the Authority, establishing partnerships with the media and relevant institutions for the development of relevant communications and public relations strategies, and executing communications and public education activities in the arrangement with the USNRC.

The NRA is developing a Strategic Communications Plan, International Relations Policy and Emergency Communications Plan. A Stakeholder Mapping and Engagement Committee has been formed with representation from the Directorates of the NRA to identify the relevant interested parties of the Authority and for the nuclear power plant and planning activities that would engage them accordingly. A skeletal media crop has been established following a media workshop organized in April 2017. Through the INSC project, the NRA communication documents were reviewed, and an emergency communication programme and procedures have been developed. Staff of the NRA have participated in public events related to nuclear power and a public event is being planned in Ghana to demonstrate the readiness of the NRA to effectively engage with the public.

III-4.2.2.   Challenges faced and lessons learned

The NRA is working with the GNPPO Public Information Group to inform the public about the nuclear power programme, including regulatory issues. Media engagements are being used to educate the public on radiation matters and the role of the NRA in the nuclear power programme. The NRA also participates in public engagement activities of the owner/operator.

III-4.3. Leadership and management for safety

III-4.3.1.   Implementation

The NRA is documenting the processes and procedures in use at the Authority to form the base of an integrated management system. Support has been received through training by the IAEA and INSC. The NRA is guided by GSR Part 2 [III–5] requirements and relies on GS-G-3.1 [III–8], GSG-12 [III–7] and IAEA Safety Standards No. GSG-13, Functions and Processes of the Regulatory Body for Safety [III–21] in the development of the integrated management system and is also seeking to comply with ISO 9001:2015 [III–6] for quality management . The management system of the NRA is being developed by building on some inherited practices (administrative and regulatory procedures) from the previous regulatory body (RPB/GAEC). It was also inspired with ideas taken from regulatory bodies from countries such as Canada, Egypt, Morocco, Lithuania, the Kingdom of the Netherlands, Pakistan and Slovenia. Information from these countries was obtained from presentations at workshops and copies of some of their management system manuals made available to the NRA.

A team of seven with representation from all the Directorates of the NRA is leading the development the Integrated Management System Manual for the NRA. The management processes being developed are directing and managing the organization; policy making; project management, process management; performance management; planning; and management of organizational change. The core processes are development of regulations and guidelines; notification and authorization; review and assessment; emergency preparedness and response; inspection; enforcement; nuclear safeguards reporting; and communication and consultation with interested parties. The support processes being developed are human resources management; knowledge management; procurement; instrumentation; estate and security management; internal audit; legal; transport management; financing/accounting; document and records control; and information and communication technology administration. Among the procedures being developed is the assessment of leadership for safety which sets out the methodology for evaluating the implementation of principles of leadership for safety of the NRA. The procedure is a component of the performance management process of the NRA. It covers all the formal and informal measures adopted by the NRA to monitor, assess and improve the leadership for safety of the Authority. The procedure covers both self-assessment and independent assessment and the identification and implementation of corrective actions.

Following the inception meeting of the INSC project in December 2019, the existing elements of the NRA management system were shared with INSC experts. A self-assessment was conducted based on GSR Part 2 with support from INSC experts [III–5]. A virtual expert visit (due to COVID-19 travel restrictions) was hosted from 22 to 26 February 2021 to discuss the existing IMS and discuss findings of the IMS review and assessment by the INSC experts. The documentation status review and assessment were conducted from 28 February to 4 March 2022 to support the drafting and revision of some processes. An on-site-assistance was held from 22 to 26 August 2022 to support in implementation of the development plan for the drafting and revision of internal regulatory procedures. A training on management system internal audit was held from 24 to 28 October 2022. The internal audit of the integrated management system in line with ISO 9001:2015 is scheduled for 17–21 April 2023 with support from INSC experts [III–6].

III-4.3.2.   *Challenges faced and lessons learned*

Challenge: With the development of integrated management system being a new thing to the NRA, it has been difficult to get staff to document processes and procedures alongside the routine activities of the Authority. Some processes and procedures need skill sets which were not present at the NRA and extra efforts by existing staff were needed to develop such processes.

Solution: Support being received from the European experts involved in the INSC project, which has assisted in the training on internal audit and in review of draft processes and procedures.

III-4.4. Development of human resources

III-4.4.1.   *Implementation*

The Ghana Atomic Energy Commission in collaboration with the University of Ghana, and support from the IAEA, established the School of Nuclear and Allied Sciences (SNAS) in September 2006 to provide education in nuclear science and technology at the graduate level. The school offers courses in nuclear engineering, computational nuclear sciences and engineering, radiation protection, applied nuclear physics, nuclear and environmental chemistry, among others. As of 2021, the school had graduated 31 doctoral candidates, 402 master of philosophy candidates and 161 IAEA postgraduate education course in radiation protection and safety of radiation sources candidates. The school is providing a pool of knowledgeable personnel for involvement in regulatory oversight of the nuclear power programme. Other courses at the various universities are providing education in other areas of need for performing nuclear regulatory activities. The graduates from SNAS and other universities have been engaged in the NRA.

The conversion of the Ghana research reactor from high enriched uranium (HEU) fuel to low enriched uranium (LEU) fuel provided a vital opportunity to demonstrate regulatory oversight of modification and refuelling of the reactor. Regulatory requirements were developed with assistance from the IAEA through expert missions.

Two expert missions and a consultancy meeting were hosted by the Ghana Atomic Energy Commission to strengthen preparations for the reactor core conversion.

In the frame of the technical cooperation project RAF4022, the first IAEA mission was conducted from 29 July to 2 August 2013 to review the safety of the core conversion from HEU to LEU fuel of the Ghana Research Reactor-1 (GHARR-1). The mission was conducted by a team consisting of an IAEA staff member (RRSS/NSNI/IAEA) and three IAEA external experts from France, South Africa, and the Syrian Arab Republic. The objective of the mission was to review the safety aspects related to the core conversion project.

The review performed by the mission team covered different safety aspects including the management system, safety analysis, operational limits and conditions, conduct of operations, technical issues and modifications and operational radiation protection.

The second mission, also within the work plan of the EC-Component of the RAF4022, Task 1.4, was conducted from 14 to 18 July 2014. The mission was conducted by three experts from the IAEA, Hungary and Canada. Topics discussed during the expert mission included: licensing process, licensing approval hold points, HEU loading procedure, inspection of SSCs, LEU core loading, commission programme, review of criticality results and emergency planning. Both the operating organization and the regulatory body were represented. Regulators of the Russian Federation, China and Ghana met in the Russian Federation in October 2014 to discuss the necessary licensing for transport of the HEU fuel. Staff of the NRA witnessed the removal and transport of spent fuel from Uzbekistan in March–April 2015. Timelines were assigned to the submittals associated with the requirements and communicated to the authorized party. The implementation of the core conversion took place in four phases: core removal, core packaging, core loading and HEU package transport phases.

The irradiated HEU fuel was to be stored for 40 days after which it was to be transported to China. Unfortunately, the necessary permissions could not be obtained from the Chinese counterparts which altered the schedule as discussed below. The transport equipment for the fuel was received in Ghana on schedule. The irradiated HEU fuel was transferred from the interim transfer cask into the licensed TUK/MNSR-C cask which was detached from its trailer in October 2016.

The activities of the LEU core loading phase were executed in five sections, namely preparation works; criticality experiments; reactivity adjustments and final reactivity determination; reactor power calibration; and commissioning.

The HEU package transport phase was supervised by NRA in collaboration with the National Security Council with active involvement of the Nuclear Security Committee on 26 and 27 August 2017.

The requirements utilized for the regulatory oversight of the core conversion hinged on the International Atomic Energy Agency (IAEA) safety standards and additional requirements issued for Minimum Qualifications required of Personnel for the Conversion Activity, Equipment Qualification and Licensing and Air Flight Requirements. The NRA received assistance from the IAEA in setting out the regulatory requirements and documents to be submitted to enable effective regulatory oversight of the core conversion. The IAEA offered to provide additional assistance if that was needed. Participation of staff in various activities of the IAEA enabled effective implementation of Ghana's mandate for the conversion.

The requirements placed on the authorized party, GAEC, included submission of the core conversion safety analysis report, schedule for the conversion activities, refurbishment plan for reactor building and reactor instrumentation, codes and standards applied, equipment certification or qualification, transportation arrangements for fuel, fuel storage plan, package design safety report, emergency preparedness and response programme, radiation protection programme, LEU core loading procedure, LEU core loading report, security plan and commissioning report. The remaining requirements were submission of test programmes for low power and power ascension tests, approach to criticality preparations, and applications including those for permits and authorization to offload HEU core and load LEU fuel.

The submissions required for issuing an operating licence included commissioning report, final operational limits and conditions, operating procedures, operational radiation protection programme, revised safety analysis report, emergency preparedness and response programme, decommissioning plan, site security plan, revised design information questionnaire and completion of an application for authorisation to operate the facility.

During the HEU package transport phase, the NRA verified submittals for safety, security, and safeguards to ensure compliance of the authorized party with national and international requirements for the transport of radioactive materials.

The NRA has developed a training programme aligned with four -quadrant model of competence as recommended in IAEA Safety Report Series No. 79, Managing Regulatory Body Competence [III–21]. Level I captures the Basic Professional Training Course (BPTC) of the IAEA, fundamentals of safeguards and non-proliferation and fundamentals of nuclear security modules. Level II addresses technical competencies in nuclear safety, nuclear security and safeguards and non-proliferation while Level III addresses the core functions of the NRA in inspections, enforcement, review and assessment and development of regulatory tools. Level IV addresses leadership and management training. Training activities have been organized from 2016 with support from IAEA, International Nuclear Safeguards Engagement Programme (INSEP) of USDOE, International Nuclear Security (INS) of USDOE, Office of Radiological Security (ORS) of USDOE, Nuclear Smuggling Detection and Deterrence (NSDD) of USDOE, ENSTTI and USNRC.

In November and December 2021, the NRA conducted the IAEA BPTC with local experts of the NRA addressing all the components of the Course. This training formed part of level training for the newly recruited staff of the NRA. In the scope of the INSC project with the European Union, the NRA has benefitted from trainings in the development of regulatory strategy, integrated management system, human resource development, development of regulations and guidelines, siting oversight and interested parties' involvement in regulatory decision-making process. Two staff benefited from on-the-job training in review and assessment of design submittals while another two received similar training on siting oversight in Hungary and Slovakia from 5 September to 28 October 2022. A staff member is currently undergoing training under the ELSE Project.

Staff of the NRA have continued to participate in IAEA trainings, technical meetings and consultancy meetings in various topical areas to develop capacity to support the regulatory oversight of the upcoming nuclear power project. A team has been constituted to evaluate the current capacities of the NRA and recommend areas where external support will be needed to support regulatory activities in Ghana.

III-4.4.2.   Challenges faced and lessons learned

The fact that the regulatory body has established a variety of tools to manage their rapid growth and has adopted innovative approaches to building a healthy organizational culture was recognized as an area of good performance at the Seventh Review Meeting of the Convention on Nuclear Safety in 2017 [III-4].

The departure of trained staff from the NRA on short notice is a challenge noticed. The NRA is considering instituting a bonding system to address this challenge.

Contribution to Programme Comprehensive Report

The NRA provided information on the regulatory infrastructure of its nuclear power programme, the impact of national laws on the nuclear power programme, requirements for safety and safeguards, requirements in implementing nuclear fuel cycle and radioactive waste management of a nuclear power plant, human resource development and needs, requirements for siting of a nuclear power project, radiation protection, public engagement, nuclear security requirements, requirements for environmental considerations for a nuclear power project; licensing regime and procedure, review of existing emergency preparedness plans and requirements for emergency preparedness for nuclear industry.

The NRA assigned the Heads of Nuclear Safety, Nuclear Security and Safeguards and Non-Proliferation to support the national team which developed the PCR. The report was submitted, and the Government announced the decision in August 2022 to deploy a nuclear power plant for electricity generation in Ghana.

Challenges faced and lessons learned

The assigned officers of the NRA were not actively involved in finalizing the PCR prior to submission to the Government.

REFERENCES TO ANNEX III

[III–1]   INTERNATIONAL ATOMIC ENERGY AGENCY, Mission Report on the Integrated Nuclear Infrastructure Review Mission Phase 1 to Ghana, 16 to 23 January 2017. https://www.iaea.org/sites/default/files/documents/review-missions/inir-mission-to-ghana-january-2017.pdf

[III–2]   INTERNATIONAL ATOMIC ENERGY AGENCY, Report of the Phase 1 Follow-Up INIR Mission to Ghana, 21–24 October 2019, Issue 10: Human Resources

Development. https://www.iaea.org/sites/default/files/documents/review-missions/inir-mission-to-ghana-january-2019.pdf

[III–3] INTERNATIONAL ATOMIC ENERGY AGENCY, Milestones in the Development of a National Infrastructure for Nuclear Power, IAEA Nuclear Energy Series No. NG-G-3.1 (Rev. 2), IAEA, Vienna (2024), https://doi.org/10.61092/iaea.zjau-e8cs

[III–4] INTERNATIONAL ATOMIC ENERGY AGENCY, The Republic of Ghana, seventh review meeting of the convention on nuclear safety national report presented by the republic of ghana in compliance with the convention on nuclear safety obligation (2017). https://www.iaea.org/sites/default/files/ghana_cns_nr_7th_2017.pdf

[III–5] INTERNATIONAL ATOMIC ENERGY AGENCY, Leadership and Management for Safety, IAEA Safety Standards Series No. GSR Part 2, IAEA, Vienna (2016).

[III–6] INTERNATIONAL ORGANIZATION FOR STANDARDIZATION, Quality management systems — Requirements, ISO 9001:2015, Geneva (2015)

[III–7] INTERNATIONAL ATOMIC ENERGY AGENCY, Organization, Management and Staffing of the Regulatory Body for Safety, IAEA Safety Standards Series No. GSG-12, IAEA, Vienna (2018).

[III–8] INTERNATIONAL ATOMIC ENERGY AGENCY, Application of the Management System for Facilities and Activities, IAEA Safety Standards Series No. GS-G-3.1, IAEA, Vienna (2006)

[III–9] INTERNATIONAL ATOMIC ENERGY AGENCY, Governmental, Legal and Regulatory Framework for Safety, IAEA Safety Standards Series No. GSR Part 1 (Rev. 1), IAEA, Vienna (2016).

[III–10] INTERNATIONAL ATOMIC ENERGY AGENCY, Establishing the Safety Infrastructure for a Nuclear Power Programme, Safety Standards Series No. SSG-16 (Rev. 1), IAEA, Vienna (2020). Convention on Nuclear Safety, INFCIRC/449, IAEA, Vienna (1994).

[III–11] Joint Convention on the Safety of Spent Fuel Management and on the Safety of Radioactive Waste Management, INFCIRC/546, IAEA, Vienna (1997).

[III–12] Convention on Assistance in the Case of a Nuclear Accident or Radiological Emergency, INFCIRC/336, IAEA, Vienna (1986).

[III–13] Convention on Early Notification of a Nuclear Accident, INFCIRC/335, IAEA, Vienna (1986)

[III–14] Convention on Supplementary Compensation for Nuclear Damage (INFCIRC/567), IAEA, Vienna (1997)

[III–15] International Convention for the Suppression of Acts of Nuclear Terrorism (ICSANT), UNODC, Vienna (2007)

[III–16] INTERNATIONAL ATOMIC ENERGY AGENCY, Code of conduct on the safety and security of Radioactive Sources, IAEA, Vienna (2004)

[III–17] INTERNATIONAL ATOMIC ENERGY AGENCY, Code of conduct on the Safety of Research Reactors, IAEA, Vienna (2004)

[III–18] INTERNATIONAL ATOMIC ENERGY AGENCY, Guidance on the Import and Export of Radioactive Sources, IAEA, Vienna (2012)

[III–19] INTERNATIONAL ATOMIC ENERGY AGENCY, The Republic of Ghana National Report for the Joint Convention on the Safety of Spent Fuel Management and on the Safety of Radioactive Waste Management (2020). https://www.iaea.org/sites/default/files/ghana-7rm.pdf

[III–20] INTERNATIONAL ATOMIC ENERGY AGENCY, Functions and Processes of the Regulatory Body for Safety, IAEA Safety Standards Series No. GSG-13, IAEA, Vienna (2018)

[III–21] INTERNATIONAL ATOMIC ENERGY AGENCY, Managing Regulatory Body Competence, Safety Reports Series No. 79, IAEA, Vienna (2014).

ANNEX IV.
CASE STUDY ON NIGERIA

IV-1. EXISTING SAFETY INFRASTRUCTURE PRIOR TO INITIATING A NUCLEAR POWER PROGRAMME IN NIGERIA

The Nigerian Nuclear Regulatory Authority (NNRA) is established by the Nuclear Safety and Radiation Protection Act (No. 19 of 1995) to regulate nuclear safety and radiological protection in Nigeria. Based on the Act, the NNRA is an independent regulatory body charged with the responsibility for nuclear safety and radiological protection regulations in the country [IV–1].

As the only Nuclear Regulatory Authority in Nigeria, the scope of work for the NNRA includes all facilities and activities involving the use of ionizing radiation and nuclear material. Consequently, the NNRA is mandated to develop regulatory infrastructure for nuclear safety in Nigeria. Following this mandate, the NNRA is implementing a regulatory control programme that includes the development of regulations and guides, authorization, oversight functions (inspection and enforcement), ancillary functions (such as training and research), and emergency preparedness and response at various stages of development throughout the lifetime of a nuclear power plant/research reactor.

The NNRA has been regulating radiation and nuclear facilities and activities since it became operational in 2001. This includes industrial, medical and agricultural applications and research reactors.

Experience in regulating the Nigerian research reactor NiRR-1

The NNRA provides regulatory oversight of the Nigerian Research Reactor (NiRR-1) which covers periodic inspections, reviews of various documents (such as the final safety analysis report) and applications (including operation and operator's licence application). The first operation licence for NiRR-1 was issued by the NNRA in 2004 when the reactor became critical. NiRR-1 is a miniature neutron source reactor (MNSR) and initially utilized high enriched uranium (HEU) fuel. The core of the reactor was converted from utilizing HEU fuel to low enriched uranium (LEU) fuel in 2018. The conversion was carried out based on the NNRA regulatory requirements and supervision. The NNRA has acquired over eighteen years of experience in regulating a MNSR. These years of experience will serve as a catalyst for regulating a multipurpose research reactor and Nuclear Power Plant.

The NNRA has established a licensing process for research reactors in Nigeria. The licensing process specifies the stages and holds point in the licensing of research reactors in Nigeria. It also specifies the relevant regulatory documents and requirements for each stage in the licensing process. A guidance document on the licensing process has been developed. The document is to provide guidance on the basis of a licensing process to be applied by the NNRA for granting licences for research reactors and their activities, including some aspects of regulatory control. The document is undergoing finalization by the NNRA.

IV-2. STATUS OF NUCLEAR POWER PROGRAMME IN NIGERIA

Nigeria has declared its intention of embarking on its first nuclear power programme and with this assertion, indicated its readiness for the peaceful application of nuclear energy for power generation. Nigeria, in its quest for a nuclear power programme, is developing infrastructure including building national institutions, establishing a legal and regulatory framework, developing human resources and financial strategies, and addressing radioactive waste management while involving stakeholders. These activities will facilitate the construction of the first nuclear power plant by adopting the IAEA Milestones approach [IV–2], which includes the 19 nuclear infrastructure elements. Additionally, it covers regulatory infrastructure for regulating a multi-purpose research reactor.

It is coordinated through the Nuclear Energy Programme Implementation Committee (NEPIC) to achieve the 19 infrastructure elements for establishing a nuclear power programme. Efforts by the Nigeria Atomic Energy Commission (NAEC) in developing Nigeria's first nuclear power plant have yielded two site locations for detailed characterization.

The NNRA is mandated to develop regulatory infrastructure for nuclear power plants in Nigeria. In accordance with this mandate, the NNRA is continuously implementing a regulatory control programme that includes the development of regulations and guides, authorization, oversight functions (inspection and enforcement), ancillary functions (such as training and research), and emergency preparedness and response amongst others.

IV-3. ROLES AND RESPONSIBILITIES OF THE GOVERNMENT AND THE REGULATORY BODY IN NIGERIA IN PHASE 1 OF THE NUCLEAR POWER PROGRAMME

IV-3.1. National policy and strategy for safety

IV-3.1.1.   Implementation

A coordinated approach to research, training and development in the areas of nuclear science and technology in Nigeria started in 1976 when the Act establishing the Nigeria Atomic Energy Commission was enacted, establishing two nuclear energy research centres in 1978. Another nuclear science and technology centre was also established in 1993, and new ones have recently been added. Consequently, trained personnel in nuclear science and technology are concentrated in these centres and some universities in Nigeria.

For the purpose of energy planning, the Energy Commission of Nigeria (ECN) was created by Decree 62 of 1979 [IV–1]. It was mandated to enunciate the national energy policy, considering the national energy needs, the available resources and how best to exploit them on a long-term basis to meet the national needs.

The Federal Government, in its effort to improve the energy generation in the country, inaugurated an Inter-Ministerial Committee on Energy Resources in April 2004. This committee assessed and quantified all the primary energy resources in the country. The study documented the various resources; natural availability, estimated derivable electricity, level of exploitation and business opportunities. Nuclear energy was identified as a significant potential source and was recommended for consideration by the Federal Government of Nigeria. Also, this recommendation was further affirmed through analytical studies using appropriate modelling tools by the ECN in 2006.

The National Energy Policy and Strategy 2003 did not comprehensively cover the nuclear power plant. Upon declaration to introduce nuclear power into the energy mix, the National Energy Policy was reviewed to consider nuclear energy for peaceful use and generation of electricity in 2013. However, the revised 2013 Energy Policy did not pay attention to nuclear safety, security and safeguards issues [IV–3].

Consequently, the Energy Policy was further reviewed and approved by the Government in 2022. The approved Energy Policy of 2022 clearly specified the objective, and the short-term, mid-term and long-term strategy for its implementation. During revision of the policy and strategy document, the NNRA participated to provide input and comment to ensure the document paid adequate attention to safety, security and safeguards issues as well as meet the fundamental safety objective and safety principles.

In furtherance to the above and to compliment the Revised National Energy Policy and Strategy of 2022, the NNRA with the collaboration of European Commission- International Nuclear Safety Cooperation (EC-INSC) is currently working on developing a National Policy and Strategy for Safety to adequately account for the fundamental safety objectives and the fundamental safety principles identified in IAEA Safety Standards Series No. SF-1, Fundamental Safety Principles[IV–7], as well as promotion of leadership and management for safety, including safety culture to clearly address the gap observed in the Revised National Energy Policy and Strategy, 2022.

IV-3.1.2.   Challenges faced and lessons learned

The process of developing and approving the National Policy and Strategy could take a longer time in order to ensure that interested parties are widely consulted. Therefore, it is important to start the development at an early stage for a transparent policy and strategy.

IV-3.2. Governmental and legal framework for safety

IV-3.2.1.   Implementation

The current legislative framework in Nigeria is contained primarily in the Nigeria Atomic Energy Commission Act No. 46 of 1976, and Nuclear Safety and Radiation Protection Act No. 19 of 1995 [IV–1].

Nigeria Atomic Energy Commission Act No. 46 of 1976 [IV–1].which established the NAEC for the development of atomic energy and all matters relating to the peaceful use of atomic energy. Nigeria is committed to implementing the nuclear power programme, leading to the activation of the NAEC in 2006.

The NNRA was established by the Nuclear Safety and Radiation Protection Act (No. 19 of 1995) [IV–3] to regulate nuclear safety and radiological protection. Pursuant to this Act, the NNRA is an independent regulatory body charged with the responsibility for nuclear safety and radiological protection regulations in the country. The Act has been reviewed as the Nuclear Safety, Security and Safeguards Bill to cover all aspects of nuclear power programme regulations including civil liability for nuclear damage. The bill has undergone public hearing and is awaiting passage into law by the National Assembly.

Act 19 of 1995 [IV–3] clearly the identified the regulatory responsibilities and power and functions of the NNRA and specifies who is qualified to be the Head of the Authority and also the establishment of the different NNRA departments, in addition it establish the governing board for NRRA and its members.

IV-3.2.2.   Challenges faced and lessons learned

Challenge

The approval process of the bill into law by the National Assembly (legislature) of the Nuclear Safety, Security and Safeguards Bill is taking longer than expected

Lessons learned

- There is a need to promulgate laws that establish and maintain an appropriate governmental and legal framework for safety, security, safeguards and civil liability for nuclear damage at an early stage of the nuclear power programme.
- Follow-up with the National Assembly/Parliament and ensuring they have adequate understanding of the bill could expedite the passage of the bill to become law.

IV-3.3. Regulatory infrastructure for safety

IV-3.3.1.   Implementation

The NNRA has been regulating radiation and nuclear facilities and activities since it 2001, coveringindustrial, medical and agricultural applications, environmental management and research reactors.

As Nigeria's sole nuclear regulator NNRA oversees all activities and facilities involving the use of ionizing radiation and nuclear material. Through the development of effective regulatory framework in Nigeria.

The NNRA developed a five year Strategic Plan to set overall goals for the NNRA aimed at analysing the institutional, organizational and technical elements and conditions that need to be established to provide a sound foundation for a high level of nuclear safety, security, safeguards and emergency preparedness which will include establishing an appropriate national legal and regulatory framework with a sound comprehensive nuclear Law, developing the necessary workforce, establishing transparent means for communication, ensuring international commitment and cooperation amongst others.

The plan proposes a new organizational structure in line with the provisions of the Nuclear Safety, Security and Safeguards Bill and adopted the mixed approach of the prescriptive and performance-based regulatory approaches.

The NNRA has developed the following regulations and guide for nuclear facilities and activities:
- Nigerian Basic Ionizing Radiation Regulations, 2003 - Radiation Protection (under review).
- Nigerian Safety Regulations for Licensing of Site for Nuclear Power Plants, 2021.
- Nigerian Nuclear Safeguards Regulations, 2021.
- Nigerian Physical Protection of Nuclear Materials and Facilities Regulations, 2021.
- Nigerian Safety of Research Reactors Regulations, 2021.

Furthermore, a number of documents are in various draft stages. These include:
- Nigerian Safety Regulations on Design and Construction of Nuclear Power Plants.

- Nigerian Safety Regulations on Commissioning of Nuclear Power Plants.
- Nigerian Safety Regulations on the Operation of Nuclear Power Plants.
- Nigerian Safety Regulations for Decommissioning of Nuclear Power Plants.
- Nigerian Licensing and Qualifications of Nuclear Power Plant Operators Regulations.
- Regulatory Guide for Selection of Technical and Scientific Support Organization.
- Guidance on Licensing process for Nuclear Power Plants.
- Guidance Document on the Safety of Research Reactors in Nigeria.
- Research Reactor Safety Inspections Manual.
- Guidance document on Nuclear Safeguards.

IV-3.3.2.   Challenges faced and lessons learned

Challenges

Prioritization and development of regulations and guides due to the absence of a decision on nuclear power technology (generic vs. technology specific regulations)

Lessons learned

i.   Experience in the oversight of research reactors assist in the development of regulatory infrastructure for a nuclear power programme.

ii.   In the absence of a decision on nuclear power technology, regulations can be developed based on 'technology neutral' using the corresponding IAEA standards.

IV-3.4. Other governmental considerations – global nuclear safety regime

IV-3.4.1.   Implementation

Global nuclear instruments

In its quest to actualize the nuclear power programme, Nigeria has assented to several international legal instruments covering nuclear safety, security, safeguards and emergency response and preparedness. The conventions and agreements that have been signed by the Nigerian Government include [IV–3]:

- 1968 Treaty on the Non-Proliferation of Nuclear Weapons (NPT) [IV–5].
  Voted in 1995 for the indefinite extension of the NPT.
- 1988 signed Comprehensive Safeguards Agreement [IV–5].
- 1988 signed Convention on Early Notification of a Nuclear Accident [IV–6].
- 1990 ratified the Convention on Early Notification of a Nuclear Accident [IV–6].
- 1990 signed Convention on Assistance in the Case of a Nuclear Accident or Radiological Emergency [IV–7].
- 2004 voted for UN Security Council Resolution 1540.
- 2007 Convention on Nuclear Safety [IV–8].
- 2007 Convention on the Safety of Spent Fuel Management and on the Safety of Radioactive Waste [IV–9].
- 2007 Convention on Physical Protection of Nuclear Material and its Amendment [IV–10].
- 2007 International Convention for the Suppression of Acts of Nuclear Terrorism [IV–11].

- 2007 Additional Protocol to the Safeguards Agreement [IV–12].
- 2007 Vienna Convention on Civil Liability for Nuclear Damage [IV–13].

The signing and ratifying of conventions and agreements is through the executive arm of government, whose primary function is to enforce laws, policymaking and maintain law and order. Most of the treaties signed by the executive require ratification by the legislature.

The ministries, departments and agencies of the Government in Nigeria fall under the executive arm. Each governmental department is responsible for implementing the law and policies concerning its work. The ministries departments and agencies that seek to enter an agreement on behalf of the Federal Government of Nigeria, are required to pass the agreement to the legislature for approval.

The NNRA, being a government parastatal, when it identifies the need to sign and ratify any convention that will be beneficial to Nigeria, goes through the Ministry of Foreign Affairs, which spearheads the process of signing such treaties and conventions.

Furthermore, the Ministry of Foreign Affairs communicates with Ministry of Justice for guidance to ensure the treaties and conventions do not contradict any national law. Upon the signing of a treaty or convention it is forwarded to the legislature for ratification. The NNRA incorporates the relevant section into its laws and regulations.

Convention on Nuclear Safety

With Nigeria's expression of political commitment in 2005 to harness nuclear energy for electricity generation, there arose the need for the country to demonstrate the peaceful and transparent nature of its nuclear power programme. Thus, in July 2005, the NNRA in conjunction with the Ministry of Foreign Affairs and the Federal Ministry of Science and Technology organized the first National Seminar on Nuclear Non-Proliferation Treaty - Challenges and Opportunities. One of the recommendations of the National Seminar was for Nigeria to, as a matter of priority, ratify the Convention on Nuclear Safety, amongst others. Consequently, Nigeria ratified the Convention in 2007.

International cooperation

International cooperation and assistance provide an opportunity to share and benefit from the experience of States that have already implemented, or are in the process of implementing, a nuclear power programme. The Nuclear Safety and Radiation Protection Act, empowered the NNRA to perform all necessary functions to enable Nigeria to meet its national and international safeguards and safety obligations in the application of nuclear energy and ionizing radiation, advice the Federal Government on nuclear security, safety and radiation protection matters; and liaise with and foster cooperation with international and other organizations or bodies having similar objectives under Section 4-(1)(d) (e)((f) of the Act [IV-1].

The NNRA signed an Arrangement for Exchange of Technical Information and cooperation in Nuclear Safety and Radiation Protection Matters [IV-1] with some competent nuclear regulatory bodies including United States Nuclear Regulatory Commission (USNRC), Rostekhnadzor of the Russian Federation, and Pakistan Nuclear Regulatory Authority (PNRA).

The NNRA has concluded an arrangement on the cooperation under the International Nuclear Safety Cooperation (INSC) programme of the European Commission to address key challenges from the IRRS mission report conducted in 2017 [VI–3]. Nigeria/NNRA became a member of the Regulatory Cooperation Forum (RCF) of the IAEA in 2012. The NNRA is also discussing cooperation with of the nuclear regulatory bodies in Belarus, Poland, the Kingdom of the Netherlands and China.

Knowledge network

Knowledge networks support the NNRA through the sharing of regulatory knowledge, practices and information and by fostering collaboration on nuclear safety, security and safeguards matters. Consequently, the NNRA belongs to the following knowledge networks to enhance competence and sharing of information:

- Forum of Nuclear Regulatory Bodies in Africa (FNRBA).
- African Commission on Nuclear Energy (AFCONE).
- Regulatory Cooperation Forum (RCF).
- Analytical Laboratories for the Measurement of Environmental Radioactivity (ALMERA).
- Global Nuclear Safety and Security Network (GNSSN).

VI–3.1.1.   Challenges faced and lessons learned

Challenges Formalizing cooperation with key regulatory bodies of countries with experience in licensing NPPs could be challenging in absence of a decision on nuclear power technology.

Lessons Learned
- Non-involvement of the National Assembly (Parliament) in the negotiation of treaties and conventions largely affects the effective implementation of treaties and conventions.
- International cooperation, including bilateral and multilateral arrangements, is an effective way to support the development of a national regulatory framework.

IV-4. KEY PRIORITIES FOR THE GOVERNMENT AND THE REGULATORY BODY IN NIGERIA IN PHASE 1

IV-4.1. Site survey

IV-4.1.1.   Implementation

Nigeria is committed to implementing the nuclear power programme, leading to the activation of the NAEC in 2006, and is mandated to promote the development of atomic energy and all matters relating to the peaceful use of atomic energy. The national institutional framework for nuclear power development in Nigeria comprises NAEC and other relevant stakeholder institutions, including NNRA. It is coordinated through the Nuclear Energy Programme Implementation Committee (NEPIC) to achieve the 19 infrastructure elements for establishing a nuclear power programme. Efforts by the NAEC in developing Nigeria's first NPP have yielded two site locations for detailed characterization from the outcome of the preliminary site assessment of the initial seven sites in different locations. These sites are also considered to have capability to accommodate units of NPPs.

During the period of preliminary site selection, the NNRA acquired good knowledge of what it takes to promulgate a law and regulations and subsequently began activities that will assist in licensing the sites. These activities include:

- Developing the competence of some senior staff in siting and site evaluation and also in the licensing process.
- Developing guidance on the licensing process for nuclear power plants in Nigeria based on IAEA Safety Standards Series No. SSG-12, Organization, Management and Staffing of the Regulatory Body for Safety [IV–14] and experience gained from the research reactor.
- Developing and issuance of Nigerian Safety Regulations for Licensing of Site for Nuclear Power Plants [IV–3].

IV-4.1.2. *Challenges faced and lessons learned*

The regulatory body needs to have detailed information of the activities of the responsible organization for the site selection to enable the regulatory body to prepare for the regulatory oversight of the site licensing.

The NNRA have issued Nigerian Safety Regulations for Licensing of Site for Nuclear Power Plants in 2021 and is continuously developing competence in regulatory oversight for effective licensing of an NPP site [IV–3]..

IV-4.2. Consultation with interested parties

IV-4.2.1. *Implementation*

Section 6(g) of the Nuclear Safety and Radiation Protection Act [IV–3]. empowers the NNRA to provide training, information and guidance on nuclear safety and radiation protection.

In order to create positive relationships with interested parties through the appropriate management of their expectations and agreed objectives, the NNRA categorizes interested parties and determines their influence and interest. It then establishes a communication management plan geared towards influencing and engaging them.

The NNRA developed a Communication Policy and Strategy which was approved by management in 2021. The effective implementation of the Policy will ensure that NNRA's intended message is understood by its target audience and the desired response achieved.

The Policy covers both external and internal means of communication between the NNRA, and the interested parties as well as the medium for monitoring and evaluation of the effects of such communication.

The internal communication of the Authority is targeted at all NNRA staff and members of the governing board for the achievement of its overall goals. It is aimed at strengthening the organizational culture and staff commitment to promote safety and security culture. While the external communication promotes knowledge and awareness of the vision, mission, mandates, activities, and services of the NNRA in relation to interested parties.

In furtherance to the above, the NNRA conducts routine workshops and training for the news media to inform them about the possible radiation risks associated with facilities and activities, and about the processes and decisions of the regulatory body. Also, the NNRA disseminates information on nuclear issues in schools to encourage current and future nuclear sector professionals. All regulations and law developed are always available for stakeholder input and comments before they are finally approved to become law.

IV-4.2.2.   Challenges faced and lessons learned

The NNRA regularly conducts workshops and training for the news media to inform them about the processes and decisions of the regulatory body. Also, the NNRA disseminate information on nuclear energy in schools to inspire current and future nuclear sector professionals.

The NNRA always make its regulations and laws available for stakeholder for input and comments before they are approved.

IV-4.3. Leadership and management for safety

IV-4.3.1.   Implementation

Prior to the IRRS mission, the NNRA had developed a management system manual. However, the mission noted that the manual should be developed in an integrated manner and all existing processes should be documented.

Consequently, the NNRA using its IAEA knowledge networks such as AFRA, FNRBA and RCF, and a technical cooperation project to develop competence in integrated management systems (IMS) through regional training, workshops and peer review.

A team was constituted to review the Integrated Management System Manual (IMSM), with representation from each department and some strategic divisions and units. The team was tasked with reviewing the manual in line with IAEA Safety Standards Series No. GSR Part 2, Leadership and Management for Safety [IV- 15] to integrate elements of safety, security, safeguards, occupational health, environment, optimization, quality, and other binding regulatory requirements, including the documentation of processes, procedures and related documents.

The IMSM described all core and support activities, management commitment and processes for improving NNRA regulatory activities. The NNRA uses a process approach to carry out or conduct its activities. These processes cover all activities and cut across the functional areas of the NNRA as identified in the IMSM. The processes are grouped into three categories: management, core and support processes and procedures.

Act 19 of 1995 [IV–1] clearly states the power and functions of the NNRA and specifies who is qualified to be the Head of the Authority and the establishment of departments including members of the board. The NNRA IMS and condition of service clearly identify the processes and procedure of selecting senior managers to ensure effective management with due consideration to safety. The senior managers are required to have the highest level of competence in nuclear science and technology, relevant laws, project management, public administrations, and other relevant discipline as it is related to their job functions including appropriate experience and decision making.

The IMS document is currently undergoing peer review with the European Commission through the INSC project arrangement with NNRA to ensure it meet the ISO 9001:2015 [IV–16] as well as GSR Part 2 [IV- 15] .

IV-4.3.2.   Challenges faced and lessons learned

The development of an IMS was difficult at the beginning. However, the support received from the IAEA and its knowledge network assisted in the development of the IMSM and documentation of the processes and procedures. The European Commission through its INSC project is assisting in the review of IMS including processes and procedures to meet ISO 9001:2015 and GSR Part 2 [IV–15] [IV–16].

IV-4.4. Development of human resources

IV-4.4.1.   Implementation

Human resource development (HRD) planning involves evaluating the current capacity of the regulatory authority, identifying gaps in meeting necessary skills, and developing strategies to close those gaps. In addressing these issues, the NNRA has developed a framework to manage the training, development, and maintenance of its staff's competence to regulate Nigeria's first NPP; and has adopted the Systematic Assessment of Regulatory Competency Needs (SARCoN) model to conduct competence need assessment on quadrant areas and specific knowledge, skills, and attitudes (KSAs) based on the structure of the NNRA including the core activities.

The NNRA conducted a Training Needs Assessment for nuclear safety with competency gap analysis for junior, intermediate and senior staff level positions in respect of the core mandate of the Authority. This was carried out using the IAEA-TECDOC-1254[6] and the IAEA SARCoN tool 0. The was later reviewed upon completion of an IAEA Workshop on Competency Development in Nuclear Safety, Security and Safeguards for regulating NPPs.

The NNRA applied the IAEA Safety Reports Series No. 79 [IV–17], Managing Regulatory Body Competence and IAEA SARCoN tool for competency assessment in nuclear safety, security and safeguards for regulating NPPs. This was carried out by determining the regulatory functions, tasks based on the required competencies and the associated KSAs for adequate competency mapping and management.

The functional job areas were classified based on the activities involved in developing regulations and guides, authorization, inspection and enforcement, emergency preparedness and response, nuclear security, nuclear safeguards, radiological safety, and public communications.

The competency profile was applied to the core functions of the NNRA using the IAEA self-assessment questionnaire for developing competency profiles for regulatory body. This was done at the level of the quadrant areas for the NNRA' staff. Every task has a defined level of competence for one KSA and any individual assigned to this position has to have a certain level of competence to perform all the task of this position. Therefore, the NNRA adopted the extended set of the SARCoN assessment criteria.

---

[6] INTERNATIONAL ATOMIC ENERGY AGENCY, Training the staff of the regulatory body for nuclear facilities: A competency framework, IAEA-TECDOC-1254, IAEA, Vienna (2001).

The information derived from the questionnaires was used to plot charts. This information was further used in formulating plans and strategies to bridge the identified competence gaps cutting across the three cadres of staff.

The national human resource and capacity building are being implemented by NAEC and NNRA. The two organizations have worked together to assess education and training needs in Nigeria.

- The NNRA estimated about 160 staff for a single unit of 1200 MWe and has hired enough staff within its headquarters and zonal offices that are responsible for regulatory oversight of NiRR-1 and other radiation facilities and activities. This staff's competence is to be developed in areas of expertise necessary to support the development of the legal and regulatory framework, site evaluation, design assessment, construction, regulatory oversight, and emergency preparedness and response using a systematic approach.
- The strategy adopted to fill the gap identified includes:
    - The developmental training programme which consists of a combination of self-study, formal training courses, workshops, Education and fellowship organized by NNRA, academic or professional organizations, regulatory bodies of other countries or by the IAEA and on the job training in the country or abroad.
    - Provision for continuing training to provide operating experience feedback from events that have occurred in the nuclear community.
    - Consideration in recruiting experienced personnel from appropriate national industries, such as the thermal power plants, the process/production industries and the oil and gas industries, and the nuclear research centre who already have some of the required competencies and may require little training to work in a nuclear regulatory body. These personnel can be a driving force in the development of the core national staff through coaching/mentoring and training.
    - Review of staff remuneration and welfare packages to commensurate industry standards, encourage mentoring, coaching, succession planning and robust condition of service.

In addition, to address the issue of nuclear education, the NAEC, as the main driver of the nuclear power programme, developed a detailed curricula for B.Sc. programmes in the sciences (with nuclear bias); National Diploma/Higher National Diploma (ND/HND) in Nuclear Science Laboratory Technology; B.Eng. Nuclear Engineering; ND/HND Nuclear Engineering Technology; three-month intensive Bridging Programme; as well as M.Sc. Nuclear Engineering/Nuclear Science. Both the three-month intensive bridging programme and the master's programme have already commenced. NAEC has identified nine universities and two polytechnics for the implementation of nuclear science and nuclear engineering programmes and has signed MOUs with six of the universities.

External support organizations and contractors

The five-year strategic plan assesses and identified national and international organizations to provide technical support. The following list covers most of the main sources of technical support, but is not intended to be all inclusive:

Support within NNRA:

The National Institute of Radiation Protection and Research was identified and assessed to serve as a technical support to the NNRA on matters related to radiation safety, nuclear safety and safeguards in particular matters related to the training of NNRA staff, licensees' calibration of radiation equipment, etc. whereas the Nuclear Security Centre would serve as technical support to the NNRA on matters related to nuclear security in particular matters related to training of NNRA staff and training and/or maintenance of licensees' or security/intelligence agencies physical protection equipment and nuclear security radiation detection equipment. The National Institute of Radiation Protection and Research and Nuclear Security Centre are undergoing upgrades to strengthen their operations in line with the legal requirements and the nuclear power programme.

The National Institute of Radiation Protection and Research is the technical support arm of the NNRA, and it offers training and education programmes in radiation protection. However, the training programmes of the Institute do not adequately address the identified training needs for safety, security and safeguards. The NNRA leverages collaboration with different international organizations for training and education purposes and to fill the identified gap. These organizations include the IAEA, USNRC, Rostekhnadzor, PNRA and EC through INSC.

The strategy for education and training also considered strengthening the National Institute for Radiation Protection and Research to carry out education, training and research needed for NPP regulation.

Domestic support (from within Nigeria):

- NNRA may appoint experts within Nigeria in the form of a Technical Advisory Committee as provided for in Section 10 of the Nuclear Safety and Radiation Protection Act IV-1]
- Government laboratories or research centres: If experimental investigation or verification is needed, advice from such bodies can be sought.
- Legal organizations: Private or governmental legal bodies that can review the language of legal documents and assist in legal enforcement actions.
- Other governmental organizations: These may be mandated to provide input on regulatory decisions but do not have specific decision-making responsibilities.

International support (from outside Nigeria):

- International organizations: The IAEA and organizations such as the International Commission on Radiological Protection (ICRP), the OECD Nuclear Energy Agency (OECD/NEA), the World Association of Nuclear Operators (WANO), the World Health Organization (WHO) and Word Nuclear Association (WNA) can be sources of advice on specific issues. This may be provided through multilateral and/or bilateral agreements or membership of their committees or by specific contractual arrangements.
- Regulatory bodies of other countries: Foreign regulatory bodies can be consulted for input through individual contacts, international cooperation agreements or international forums on the decision-making process used in a given area, which can be particularly useful when designs or regulatory procedures utilized in one State are considered in another.

The NNRA signed an Arrangement for Exchange of Technical Information and Cooperation in Nuclear Safety and Radiation Protection Matters with some competent nuclear regulatory bodies including the USNRC), Rostekhnadzor and the PNRA. The NNRA has also concluded an arrangement of cooperation under the INSC programme of the European Commission to address key challenges identified in the IRRS mission conducted in 2017.

Regulatory bodies of vendor States: regulatory bodies of vendor States can be consulted with respect to the regulatory structure and its application in a State from which structures such as the reactor pressure vessel, components or services are provided to the licensee.

IV-4.4.2.   Challenges faced and lessons learned

Challenges

- Availability of skilled personnel to perform the regulatory functions for the nuclear power programme.
- Rules and regulations on recruitment and employment in Government.
- Retention of qualified staff and succession planning for key managerial positions.
- Strengthening the National Institute of Radiation Protection and Research to carry out education, training and research needed for NPP regulation.

Lessons learned

- International cooperation, including bilateral and multilateral arrangements, is an effective way to support the development of a national regulatory framework. This has assisted the NNRA in capacity building.
- Establishment and development of a TSO in the NNRA will assist in capacity building.
- Reviewing staff remuneration and welfare packages to make them commensurate with the industry standard would encourage mentoring, coaching, succession planning and robust conditions of service.

REFERENCES TO ANNEX IV

[IV–1]   CONVENTION ON NUCLEAR SAFETY, NIGERIAN NATIONAL REPORT FOR THE 8 TH REVIEW MEETING (2019). https://www.iaea.org/sites/default/files/national_report_of_nigeria_for_the_8th_revie w_meeting.pdf

[IV–2]   INTERNATIONAL ATOMIC ENERGY AGENCY, Milestones in the Development of a National Infrastructure for Nuclear Power, IAEA Nuclear Energy Series No. NG-G-3.1 (Rev. 2), IAEA, Vienna (2024), https://doi.org/10.61092/iaea.zjau-e8cs

[IV–3]   INTEGRATED REGULATORY REVIEW SERVICE (IRRS) MISSION TO NIGERIA (2017). https://www.iaea.org/sites/default/files/documents/review-missions/irrs_nigeria_report_2017-10-04.pdf

[IV–4]   EUROPEAN ATOMIC ENERGY COMMUNITY, FOOD AND AGRICULTURE ORGANIZATION OF THE UNITED NATIONS, INTERNATIONAL ATOMIC ENERGY AGENCY, INTERNATIONAL LABOUR ORGANIZATION, INTERNATIONAL MARITIME ORGANIZATION, OECD NUCLEAR ENERGY AGENCY, PAN AMERICAN HEALTH ORGANIZATION, UNITED NATIONS ENVIRONMENT PROGRAMME, WORLD HEALTH ORGANIZATION, Fundamental Safety Principles, IAEA Safety Standards Series No. SF-1, IAEA, Vienna (2006), https://doi.org/10.61092/iaea.hmxn-vw0a

[IV–5] Treaty on the Non-Proliferation of Nuclear Weapons, INFCIRC/140, IAEA, Vienna (1986).

[IV–6] Convention on Early Notification of a Nuclear Accident, INFCIRC/335, IAEA, Vienna (1986).

[IV–7] Convention on Assistance in the Case of a Nuclear Accident or Radiological Emergency, INFCIRC/336, IAEA, Vienna (1986).

[IV–8] Convention on Nuclear Safety, INFCIRC/449, IAEA, Vienna (1994).

[IV–9] Joint Convention on the Safety of Spent Fuel Management and on the Safety of Radioactive Waste Management, INFCIRC/546, IAEA, Vienna (1997).

[IV–10] Convention on the Physical Protection of Nuclear Material, INFCIRC/274/Rev. 1, IAEA, Vienna (1979).

[IV–11] International Convention for the Suppression of Acts of Nuclear Terrorism (ICSANT), UNODC, Vienna (2007)

[IV–12] Additional Protocol to CSA, INFCIRC/540, IAEA, Vienna (1997).

[IV–13] Vienna Convention on Civil Liability for Nuclear Damage, INFCIRC/500, IAEA, Vienna (1963).

[IV–14] INTERNATIONAL ATOMIC ENERGY AGENCY, Organization, Management and Staffing of the Regulatory Body for Safety, IAEA Safety Standards Series No. GSG-12, IAEA, Vienna (2018).

[IV–15] INTERNATIONAL ATOMIC ENERGY AGENCY, Leadership and Management for Safety, IAEA Safety Standards Series No. GSR Part 2, IAEA, Vienna (2016).

[IV–16] INTERNATIONAL ORGANIZATION FOR STANDARDIZATION, Quality management systems — Requirements, ISO 9001:2015, Geneva (2015).

[IV–17] INTERNATIONAL ATOMIC ENERGY AGENCY, Managing Regulatory Body Competence, Safety Reports Series No. 79, IAEA, Vienna (2014).

ANNEX V.
CASE STUDY ON UNITED ARAB EMIRATES

V-1. EXISTING SAFETY INFRASTRUCTURE PRIOR TO INITIATING A NUCLEAR POWER PROGRAMME IN THE UNITED ARAB EMIRATES

Prior to the initiation of the nuclear power programme, the Federal Law No. (1) of 2002 regarding the Regulation and Control of the Use of Radiation Sources and Protection against Their Hazards, issued on 6 January 2002 was the instrument regulating the uses of nuclear applications in the United Arab Emirates (UAE) [V–1]. It established within the Ministry of Electricity and Water a department responsible for regulating and controlling the use of radiation sources, and the protection against their hazards. The Federal Law No. (1) of 2002 also provided for the creation of a Radiation Protection Committee. The Department established under the Ministry of Electricity and Water has then been shifted to the Federal Environmental Agency (FEA) in accordance with the UAE Federal Law No. (20) of 2006.

The UAE issued in April 2008 the Policy of the UAE on the Evaluation and Potential Development of Peaceful Nuclear Energy (hereinafter referred to as the UAE Nuclear Policy) [16] which identified the actions that the Government of the UAE would undertake should it choose to move forward in establishing an actual nuclear power programme. The actions of the Government would be guided by inter alia six principles and several targets, including the development of a comprehensive national legal framework covering safety, security, non-proliferation and nuclear liability.

The Federal Law by Decree No. (6) of 2009 on the Peaceful Uses of Nuclear Energy, issued on 23 September 2009, (hereinafter referred to as the Nuclear Law), sets in place the framework for nuclear regulation and formally established the nuclear regulatory body, the Federal Authority for Nuclear Regulation (FANR) as an independent organization with full legal competence and financial and administrative independence [V–3].

V-2. STATUS OF NUCLEAR POWER PROGRAMME IN THE UNITED ARAB EMIRATES

The Barakah Nuclear Power Plant (BNPP) is the first nuclear power plant in the UAE and consists of four APR-1400 nuclear reactors. The BNPP is located about 300 km west of the city of Abu Dhabi, on the shore of the Arabian Gulf.

As of the date of submission of this TECDOC, FANR has issued three licences for operation of three of the units of the BNPP, respectively in February 2020 for the operation of Unit 1, in March 2021 for the operation of Unit 2 and in June 2022 for the operation of Unit 3. The three abovementioned units of the BNPP are in commercial operation and a fourth unit is in the final stages of construction and is expected to be licensed in the near future.

The Nawah Energy Company (Nawah) is the licensee and operator of the Units 1 to 3 of the BNPP. To date, Nawah has completed the first refuelling outage for Unit 1 in July 2022, and Unit 2 in April 2023. In addition, FANR has issued in April 2023 a licence to Emirates Nuclear

Energy Company (ENEC) for the selection of a site for the construction of a radioactive waste management facility.

V-3. ROLES AND RESPONSIBILITIES OF THE GOVERNMENT AND THE REGULATORY BODY IN THE UNITED ARAB EMIRATES IN PHASE 1 OF THE NUCLEAR POWER PROGRAMME

V-3.1. National policy and strategy for safety

V-3.1.1. Implementation

The Policy of the UAE on the Evaluation and Potential Development of Peaceful Nuclear Energy (hereinafter referred to as the UAE Nuclear Policy) was adopted by the UAE Cabinet of Ministers in April 2008. The UAE Nuclear Policy outlines the role of nuclear energy in the UAE's energy programme and stated the UAE's commitment to operational transparency and to the highest standards of safety, security, and nuclear non-proliferation throughout the life of the nuclear programme. The UAE Nuclear Policy also discussed the UAE's intent to develop its peaceful domestic nuclear power capability in partnership with the governments and firms of responsible nations as well as with the assistance of appropriate expert organizations including the International Atomic Energy Agency (IAEA) in a manner that best ensures long-term sustainability.

Following the issuance of UAE Nuclear Policy, the UAE established a nuclear energy programme implementing organization (NEPIO) as recommended by the IAEA and the Executive Affairs Authority (EAA) of Abu Dhabi performed the functions of the NEPIO supporting the development of both the UAE nuclear regulatory body and the UAE nuclear owner/operator prior to its formal, legal establishment.

In September 2008, the EAA developed an internal strategy document titled 'Roadmap to Success' which built on the IAEA Milestones covering the 19 nuclear infrastructure issues and set forth the early path for the programme [V–4]. This Roadmap addressed the broad range of issues including legislation, capacity building, and radioactive waste management and established recommendations, objectives, identified the responsibilities of various interested parties in the UAE and timelines for specific targets [V–5]. The EAA continued to support the programme until the formal establishment of FANR as the independent national nuclear regulatory body by the Nuclear Law and formal establishment of ENEC by Law No. (12) of 2009.

V-3.1.2. Challenges faced and lessons learned

An Integrated Regulatory Review Service (IRRS) mission to the UAE in 2011 reviewed the effectiveness of the UAE's framework for safety and the establishment of the safety infrastructure to support the rapidly developing nuclear power programme. The IRRS mission noted that while the UAE developed its Nuclear Policy within a relatively short time frame, it was based on a firm analysis of future demand for electricity. The IRRS review team considered the Nuclear Policy and the Nuclear Law, to constitute an adequate platform for addressing the Fundamental Safety Principles .[V-6] Whilst a national policy is a requirement of IAEA Safety Standards Series No. GSR Part 1 (Rev. 1), Governmental, Legal and Regulatory Framework for Safety [V–6], the IRRS review team concluded that such policies are rarely concisely formulated and transparently communicated, even in countries with mature nuclear power programmes. The IRRS review team considered it particularly important to issue such policy statements in a country that has just started to explore the nuclear option for addressing its energy demand. The IRRS review team also noted that in implementing the Nuclear Policy through the Nuclear Law, UAE consulted widely with interested parties. The IRRS review team therefore considered the Nuclear Policy to be good practice.

V-3.2. Governmental and legal framework for safety

V-3.2.1. Implementation

The nuclear legislative framework in the UAE includes the Nuclear Law, which is a comprehensive nuclear legislation establishing FANR as the nuclear regulatory authority and addressing nuclear safety, radiation safety, security and safeguards.

It is complemented by the Federal Law by Decree No. (4) of 2012 Concerning Civil Liability for Nuclear Damage [V–3], which came into effect in August 2012 (referred to as the Nuclear Liability Law) and aims to establish the rules and requirements relating to civil liability and compensation for nuclear damage applicable in a situation of a nuclear accident occurring in a nuclear installation in the UAE or during a transport of nuclear material to or from a nuclear installation located in the UAE. The provisions of the Nuclear Liability Law are in line with UAE obligations under the 1997 Vienna Convention on Civil Liability for Nuclear Damage to which the UAE is a party and take into account best international practices [V–8].

These two legislations are complemented by the following Cabinet Resolutions [V–9]:

- Cabinet Resolution No. 8 of 2014 Concerning License Fees and Services Provided by the Federal Authority for Nuclear Regulation.
- Cabinet Resolution No. 27 of 2015 Concerning Administrative Penalties for Violating the Conditions of the Licenses issued by the Federal Authority for Nuclear Regulation. This Cabinet Resolution is soon to be replaced by the Cabinet Resolution Concerning Violations and Administrative Penalties Imposed by FANR.
- Cabinet Resolution No. 104 of 2021 Concerning the Fees for Calibration Services provided by the Secondary Standards Dosimetry Laboratory of the Federal Authority for Nuclear Regulation.
- Cabinet Resolution No. 122 of 2022 Concerning the Fees for the NORMS related Licenses and some Activities Conducted by the Federal Authority for Nuclear Regulation.

In addition to these key legislative instruments, the following other legislations are also relevant [V–9] [V–10]:

- Abu Dhabi Law No. (8) of 2021 Establishing the Emirates Nuclear Energy Company, Public Joint Stock Company (PJSC), issued on 26 August 2021 (referred to as the ENEC Law).
- Federal Law by Decree No. (2) of 2011 Concerning the Establishment of the National Emergency, Crisis and Disasters Management Authority issued on 19 July 2011, as amended.
- Federal Law No. (24) of 1999 Concerning the Protection and Development of the Environment, issued on 17 October 1999, as amended by Federal Law No. (11) of 2006.
- Law No. (14) of 2007 Concerning the Establishment of the Critical National Infrastructure Authority and Law No. (1) of 2012 Concerning the Abolishment of the

Critical National Infrastructure Authority, issued on 28 February 2012. This law transferred the functions and responsibilities of the Critical National Infrastructure Authority (CNIA) to the Critical Infrastructure and Coastal Protection Authority (CICPA) of the UAE Armed Forces.

- Federal Law No. (14) of 2016 Concerning Administrative Violations and Penalties in Federal Government.
- Federal Law No. (3) of 1987 concerning the Penal Code, as amended.

V-3.2.2.   *Challenges faced and lessons learned*

During an Integrated Nuclear Infrastructure Review (INIR) mission in 2011, the review team observed that the UAE has promulgated a Nuclear Law which covers safety, security and safeguards but not civil liability for nuclear damage. The mission team recognized that the UAE infrastructure was progressing rapidly and was well advanced, having reached Milestone 1 and all elements of Milestone 2 except for the adoption of an international instrument on civil liability for nuclear damage and promulgation of associated implementing legislation [V–4]. This recommendation from the INIR mission was closed by the UAE, which acceded in May 2012 to the 1997 Vienna Convention on Civil Liability for Nuclear Damage [V–1] [V–8]. In line with the recommendations formulated the International Expert Group on Nuclear Liability (INLEX), the UAE also joined in August 2012 the Joint Protocol relating to the Application of the Vienna Convention and the Paris Convention and the Convention on Supplementary Compensation for Nuclear Damage in July 2014. As mentioned above, the UAE issued also the Federal Law by Decree No. (4) of 2012 Concerning Civil Liability for Nuclear Damage, which came into effect in August 2012 [V–4] [V–11].

V-3.3. Regulatory infrastructure for safety

V-3.3.1.   *Implementation*

The Federal Authority for Nuclear Regulation (FANR) is established by the Nuclear Law in Article (4) as a public organization with an independent balance sheet, an independent legal personality, full legal competence, and financial and administrative independence. Article (4) (1) and (2) of Nuclear Law states that the aims of FANR is to regulate the nuclear sector in the State towards peaceful purposes only, and to ensure safety, nuclear security and radiation protection.

The Nuclear Law identifies the specific functions of FANR when exercising regulatory control and supervision of the nuclear sector, including the development and issuance of regulations and regulatory guides, the licensing of regulated activities, the conduct of inspections and the taking, as may be required, of enforcement actions. Article (6) of the Nuclear Law gives FANR exclusive jurisdiction over the licensing of regulated activities in the UAE and other activities within the scope of the Nuclear Law.

Chapter three of the Nuclear Law (Articles (10) to (17)) establishes requirements for the management of FANR, including the establishment and powers of its board of management, the appointment of the Board members by the UAE Cabinet of Ministers. Article (10) specifies that members of the Board shall not engage whether directly or indirectly in the conduct of any regulated activity under the jurisdiction of FANR and shall not have any personal interest conflicting with FANR's interests. Article (14) of the Nuclear Law addressed the designation and responsibilities of the FANR Director General.

The Nuclear Law empowers FANR to develop and issue regulations and regulatory guides, which are required for the implementation of the Nuclear Law and specifying the requirements to be complied with in the conduct of regulated activities. In developing regulations and regulatory guides, FANR takes into consideration comments from interested parties, information made available by experts, and internationally recognized standards and recommendations, such as IAEA standards.

FANR has established within its integrated management system (IMS), a systematic process for the development and review of regulations and guides that includes provisions for consultation with external interested parties, and review and incorporation of their comments. This process is complemented by a systematic review mechanism of the existing regulations and regulatory guides following a five-year cycle.

FANR has now in place a set of 24 regulations, complemented by 22 regulatory guides.

V-3.3.2. Challenges faced and lessons learned

The IRRS mission to the UAE in 2011 noted that FANR has developed a set of regulations and guides to support the regulatory review of the licence applications related to the BNPP. That set of regulations and guides draws heavily upon IAEA guidance, and foreign regulators regulations such as United States Nuclear Regulatory Commission (USNRC). The choice of the Korea Electric Power Corporation (KEPCO) design, together with the similarity of the Korean and US regulations, guides and codes, has greatly facilitated FANR's effort to establish a comprehensive regulatory framework in a relatively short time.

V-3.4. Other governmental considerations – global nuclear safety regime

V-3.4.1. Implementation

In order to meet the UAE's commitments on transparency and international cooperation as underpinned in the UAE Nuclear Policy, the UAE has acceded to the relevant international instruments in the areas of nuclear safety, nuclear security, and nuclear non-proliferation as listed below

- Convention on Nuclear Safety (CNS) acceded to by the UAE on 31 July 2009 and entered into force for the UAE on 29 October 2009 [V–12].
- Joint Convention on the Safety of Spent Fuel Management and on the Safety of Radioactive Waste Management (Joint Convention), acceded to by the UAE on 31 July 2009 and entered into force for the UAE on 29 October 2009 [V–13] [V–14].
- Convention on Early Notification of a Nuclear Accident acceded to by the UAE on 2 October 1987 and entered into force for the UAE on 2 November 1987 [V–15].
- Convention on Assistance in the Case of a Nuclear Accident or Radiological Emergency acceded to by the UAE on 2 October 1987 and entered into force for the UAE on 2 November 1987 [V–16].
- Convention on the Physical Protection of Nuclear Material acceded to by the UAE on 16 October 2003 and entered into force for the UAE on 15 November 2003 [V–17].
- Amendment to the Convention on the Physical Protection of Nuclear Material, accepted by the UAE on 31 July 2009 and entered into force on 8 May 2016 [V–18].
- Treaty on the Non-Proliferation of Nuclear Weapons (NPT), acceded to by the UAE on 26 September 1995 [V–19].

- Agreement between the United Arab Emirates and the International Atomic Energy Agency for the Application of Safeguards in Connection with the Treaty on the Non-Proliferation of nuclear weapons (Safeguards Agreement), signed on 15 December 2002 and entered into force for the UAE on 9 October 2003.
- Protocol Additional to the Agreement between the United Arab Emirates and the International Atomic Energy Agency for the Application of Safeguards in Connection with the Treaty on the Non-Proliferation of nuclear weapons (Additional Protocol) [V–20], signed on 8 April 2009 and entered into force for the UAE on 20 December 2010.
- Protocol to Amend the Vienna Convention on Civil Liability for Nuclear Damage, acceded to by the UAE on 29 May 2012 and entered into force for the UAE on 29 August 2012 [V–21].
- Joint Protocol Relating to the Application of the Vienna Convention and the Paris Convention, acceded to by the UAE on 29 August 2012 and entered into force for the UAE on 29 November 2012 [V–22].
- Convention on Supplementary Compensation for Nuclear Damage, ratified by the UAE on 7 July 2014 and entered into force for the UAE on 15 April 2015 [V–23].

In addition, the UAE has expressed its political commitment to follow the guidance as contained in the Code of Conduct on the Safety and Security of Radioactive Sources and its supplementary Guidance on the Import and Export of Radioactive Sources [V–24]. Finally, the UAE has also declared its adherence to the Nuclear Suppliers Group Guidelines and its commitment to act in accordance with such Guidelines (INFCIRC/969) [V–25].

Since it became a Contracting Party to the safety conventions (CNS and Joint Convention), the UAE has participated actively in the review processes of such international instruments. In addition, FANR is participating and contributing to the work of the committees under the IAEA Commission on Safety Standards.

IAEA peer reviews

Integrated Nuclear Infrastructure Review (INIR) mission.

In January 2011, an IAEA team of experts conducted an Integrated Nuclear Infrastructure Review (INIR) mission of the UAE's national infrastructure for the introduction of nuclear power. The review assessed the status of UAE national infrastructure for the introduction of a national nuclear power programme [V-5].

The mission team recognized that the UAE infrastructure is progressing rapidly and is well advanced and concluded that the UAE has reached Milestone 1, having made a knowledgeable decision regarding its nuclear power programme [V–4]. The mission team further concluded that the UAE has accomplished all of the conditions for Phase 2, in each of the 19 nuclear infrastructure issues, with the exception of the adoption of an international instrument on civil liability for nuclear damage and promulgation of associated implementing legislation.

In June 2018, an IAEA team of experts conducted an INIR Phase 3 mission, which is the final stage of the INIR process [V–3]. The review found that the UAE has adequately developed its nuclear infrastructure including a legal and regulatory framework and has established and staffed the required organizations in particular an independent regulatory body and an operating organization. The review also provided a number of recommendations and suggestions for further strengthening the national nuclear infrastructure in the UAE, such as promoting a proactive coordination of future programme developments at a national level; continuing review and revision of the legal and regulatory framework and establishing a national research and development programme.

Integrated Regulatory Review Service (IRRS) mission.

In December 2011, an Integrated Regulatory Review Service (IRRS) mission was conducted with the objective to review the effectiveness of the UAE framework for safety as implemented by FANR and of the safety infrastructure to support the rapidly developing nuclear power programme [V–1].

The IRRS mission team identified a number of good practices and made recommendations and suggestions indicating where improvements are necessary or desirable to continue enhancing the effectiveness of regulatory functions in line with the IAEA Safety Standards.

Integrated Regulatory Review Service (IRRS) extended follow-up mission.

An IRRS follow-up mission was conducted in January 2015 to review the measures undertaken following the recommendations and suggestions of the 2011 IRRS mission [V–26]. In addition, the follow-up mission was extended to include the transport of radioactive material.

The IRRS review team concluded that the recommendations and suggestions from the 2011 IRRS mission have been taken into account systematically by a comprehensive action plan and also provided additional recommendations and suggestions. Significant progress had been made in many areas and many improvements were carried out following the implementation of the action plan.

Emergency Preparedness Review (EPREV) Service

In March 2015, an Emergency Preparedness Review (EPREV) Service examined the UAE's progress in preparing the necessary response measures to be applied in the unlikely event of a nuclear emergency at the BNPP. It found that the UAE has established effective, nuclear emergency management arrangements that build upon the nation's existing emergency response system [V–27]. A national action plan to address the IAEA's recommendations and suggestions was prepared and implemented under the coordination of FANR and in cooperation with a number of entities with responsibilities on-site and off-site. The action plan implementation led to several improvements which were assessed during an EPREV follow-up mission in September 2019.

During the EPREV follow-up mission it was noted that UAE has made significant progress in developing and revising emergency arrangements for an emergency at BNPP since the 2015 EPREV mission. The review team noted a number of good practices related to the EPR arrangements in the country and noted areas that could benefit from further improvements.

Education and Training Appraisal (EduTA) mission

In February 2017, an Education and Training Appraisal (EduTA) was carried out under the IAEA technical cooperation project RAS/9/081 'Providing Education and Training in Radiation Safety in the Asia-Pacific Region'. The EduTA team identified several recommendations and suggestions that could further enhance the overall national capabilities and performances for education and training in radiation protection and safety.

Pre-Operational Safety Review Team (OSART) mission

In September and October 2017, an 18-day pre-operational safety review (otherwise known as a pre-OSART mission) was conducted at the Barakah site [V–28].

This OSART mission reviewed twelve areas: leadership and management for safety; training and qualification; operations; maintenance; technical support; operating experience feedback; radiation protection; chemistry; emergency preparedness and response; accident management; human technology organization interaction; and commissioning.

The review team identified a number of good practices that were shared with the nuclear industry globally and also made a number of recommendations to improve operational safety.

Bilateral cooperation agreements concluded by the United Arab Emirates

The UAE has also concluded agreements with several nations to advance cooperation in the peaceful uses of nuclear energy. The national level agreements are listed below. A number of subsidiary cooperative agreements have been reached with national regulatory bodies and other entities pursuant to these high-level agreements [V-3].

- Agreement for Cooperation between the Government of the United Arab Emirates and the Government of the Republic of France in the Development of Peaceful Uses of Nuclear Energy, 15 January 2008.
- Agreement between the Government of the United Arab Emirates and the Government of the Republic of Korea for Cooperation in the Peaceful Uses of Nuclear Energy, 22 June 2009.
- Agreement for Cooperation between the Government of the United Arab Emirates and the Government of the United States of America Concerning Peaceful Uses of Nuclear Energy, 21 May 2009.
- Agreement between the Government of the United Kingdom of Great Britain and Northern Ireland and the Government of the United Arab Emirates for Cooperation in the Peaceful Uses of Nuclear Energy, 2010.
- Agreement between the Government of the United Arab Emirates and the Government of Australia on Cooperation in the Peaceful Uses of Nuclear Energy, 31 July 2012.
- Agreement between the Government of the United Arab Emirates and the Government of Canada for Cooperation in the Peaceful Uses of Nuclear Energy, 18 September 2012.
- Agreement between the Government of the United Arab Emirates and the Government of the Russian Federation on Cooperation in the Field of the Use of Nuclear Energy for Peaceful Purposes, 17 December 2012.
- Agreement on Cooperation in the Peaceful Uses of Nuclear Energy between the United Arab Emirates and the Argentine Republic, 14 January 2013.
- Agreement between the Government of the United Arab Emirates and the Government of Japan for Cooperation in the Peaceful Uses of Nuclear Energy, 2 May 2013.
- Agreement for Cooperation on Peaceful Uses of Nuclear Energy between the Government of the Kingdom of Saudi Arabia and the Government of the United Arab Emirates in the Peaceful Uses of Nuclear Energy, 27 November 2019.

V-3.4.2. Challenges faced and lessons learned

The IRRS mission to the UAE in 2011 noted that UAE through FANR, has taken a very active role in international collaboration and contributed actively to the development of the global safety regime. The team also noted that international engagement is driven by the ambition to achieve the highest standards of performance in the nuclear sector, including safeguards and non-proliferation, as clearly laid out in the Nuclear Policy.

The IRRS mission to the UAE in 2011 also confirmed that the UAE and FANR have given evidence of ambitious use of international peer review missions as well as demonstrated that the findings from these missions are incorporated into actions plans and that resulting actions are being implemented. The effectiveness and efficiency by which this has taken place is considered good practice.

V-4. KEY PRIORITIES FOR THE GOVERNMENT AND THE REGULATORY BODY IN THE UNITED ARAB EMIRATES IN PHASE 1

V-4.1. Site survey

V-4.1.1. Implementation

The Nuclear Law gives FANR the authority to regulate the nuclear sector of the UAE. Article (25) of the Nuclear Law requires that a licence be obtained from FANR prior to engaging in any regulated activity, which includes selection of a site for construction of a nuclear facility, preparation of a site for construction of a nuclear facility, construction of a nuclear facility, commissioning and operation of a nuclear facility [V–29].

FANR has developed several regulations establishing the regulatory requirements related to site selection, site preparation and site construction.

FANR-REG-02, Regulation for the Siting of Nuclear Facilities [V–30], describes the requirements for the evaluation of a proposed site and defines the extent of information relating to a proposed site to be submitted by the applicant for a licence for the selection of a site. This information includes, but is not limited to, the following:

- Evaluating a proposed site to ensure that the site-related phenomena and characteristics are adequately taken into account.
- Analyzing the characteristics of the population of the region and the capability of implementing an emergency plan over the projected lifetime of the plant.
- Defining site-related hazards.
- Quantifying the input parameters related to seismic, meteorological, hydraulic, geotechnical areas and human induced conditions used in the design of the nuclear facility systems, structures and components (SSCs).

FANR-REG-02 specifies the requirements to be used within the UAE when evaluating potential sites for NPPs. The regulation requires the applicant/licensee to establish "measures to monitor and verify all site characteristics remain within the assumptions used in the design and the final safety analysis report throughout the life of the nuclear facility" [V–30].

FANR-REG-03, Regulation for the Design of Nuclear Power Plants, includes the design requirements for structures, systems, and components important to safety that must be met for safe operation of a nuclear facility, and for preventing or mitigating the consequences of potential events that could jeopardize safety [V–31]. It also establishes requirements for a comprehensive safety assessment, which is carried out in order to identify the potential hazards that may arise from the operation of the nuclear facility, under the various plant states. It specifies that natural external events shall be considered in the design process including those which have been identified in site characterization, such as earthquakes, dust storms/sandstorms, cyclones, floods, high winds, tornadoes, tsunami (tidal waves), and extreme meteorological conditions.

FANR-REG-06, Regulation for an Application for a Licence to Construct a Nuclear Facility, requires an applicant for a licence to construct a nuclear facility to submit in the preliminary safety analysis report (PSAR) comprehensive information on the evaluation of the proposed site [V–32].

FANR-REG-14, Regulation for an Application for a Licence to Operate a Nuclear Facility, specifies the requirements for an application to FANR for a licence for the operation of a nuclear facility, including the submission in the final safety analysis report (FSAR) of an updated evaluation of the site and the nuclear commissioning tests[V–32] .

Once the UAE-wide siting study was conducted and the final candidate sites determined, the UAE undertook site characterization studies to evaluate construction suitability and to pinpoint site-specific design criteria. In parallel, extensive environmental studies were undertaken to yield site-specific and regional data to assess environmental impacts. As the selection process occurred early on in the UAE's programme, the UAE sought IAEA guidance and requested the IAEA to evaluate the country's criteria of site selection, with the Agency's review resulting in a positive conclusion [V–29].

A site selection process was undertaken using IAEA and other international guidance materials. FANR has issued three licenses to ENEC: Licence for Selection of a Site for the Construction of a Nuclear Facility, Licence for Preparation of a Site for the Construction of a Nuclear Facility, and Limited Licence for the Construction of a Nuclear Facility.

On 17 July 2012, based on a thorough review and assessment of ENEC's construction licence application, FANR issued a licence to ENEC authorizing the construction of Units 1 and 2 at BNPP. FANR conducted a thorough review of the construction licence application for BNPP Units 1 and 2 including Chapter 2, Site Characteristics, of the PSAR. FANR concluded that Chapter 2 of the PSAR along with supplemental application materials demonstrated a sufficient safety basis for issuing a construction licence and complied with relevant regulatory requirements, contained primarily in FANR-REG-02 [V–30]. FANR concluded that the site has been properly characterized for the environment of the UAE at the location of the BNPP and that the site is suitable for use as a location for operation of a multi-unit nuclear power facility as described in the PSAR for BNPP Units 1 and 2 supplied by ENEC [V–30].

V-4.1.2.   Challenges faced and lessons learned

When Fukushima Daiichi accident occurred in Japan, FANR had just begun to evaluate the first construction license application it had received from ENEC to build new reactors at the BNPP site.

To learn from the accident at Fukushima Daichi in Japan, FANR established a Fukushima Task Force to follow and engage in activities of other regulatory bodies and international organizations, such as the IAEA, the European Nuclear Safety Regulators Group (ENSREG), the Nuclear Energy Agency of the Organisation for Economic Co-operation and Development

(OECD NEA), the US Nuclear Regulatory Commission, Korea's Nuclear Safety and Security Commission (NSSC), and the Korea Institute of Nuclear Safety (KINS).

FANR completed assessments of the UAE's nuclear safety relevant regulations and guides and determined that they sufficiently covered the possibility of unexpected extreme events.

FANR also recommended ENEC to conduct a detailed assessment of the Barakah facility using a 'stress test' methodology and submit a report on the assessment results with identification of any planned safety improvements. ENEC completed the assessment, reported the results to FANR and proposed a number of design enhancements to protect the BNPP from extreme events.

Lessons learned from the Fukushima Daiichi accident were evaluated by ENEC and Nawah and changes implemented where necessary. This evaluation followed an approach modelled on the 'stress test' approach of ENSREG.

The INIR mission to the UAE in 2018 noted that all information obtained during the construction phase by ENEC and Nawah has confirmed the original site characterisation assessment. In addition to this, Nawah had a plan for ongoing monitoring to ensure the site continues to meet the design intent as required by the regulation. This plan has been accepted by FANR. Requirements for a 10-yearly review of the site related aspects pertaining to nuclear safety have been established by FANR in the regulation REG-16 for Operational Safety including Commissioning [V–33].

V-4.2. Consultation with interested parties

V-4.2.1. Implementation

Article (9) of the Nuclear Law requires FANR to maintain the highest standards of transparency whilst performing its functions and to facilitate the public's access to all relevant information to its activities [V–1].

In order to meet its obligations for transparency and in line with the UAE Federal Government Communication Strategy, FANR has developed and implemented its communication strategy and programmes targeting different stakeholders including the public, government entities, international organizations, licensees, students, media and employees. It uses different channels and tools to engage and communicate with its stakeholders such as the FANR website (www.fanr.gov.ae), which includes thorough and detailed information on FANR's activities. The website presents the following:

- All published regulations and guides with the exception of those containing sensitive information, which are restricted for nuclear security reasons.
- Resolutions adopted by the Board of Management.
- Summaries of the safety evaluation reports.
- Summaries of inspection reports.
- Peer review reports.
- FANR Annual Reports.
- UAE National Reports on the implementation of safety-related conventions.

The FANR website allows stakeholders and the public to interact with FANR using various communication channels including:

- Live chat, e-Forum, Wasl (public and licensees' enquiry channel), Talk to the Director General', and email queries to FANR.

In addition, consultation with stakeholders is embedded in the FANR process for the development and revision of FANR regulations and regulatory guides. Draft regulations and guides are shared on FANR website and communicated to identified stakeholders and the inputs received are assessed and taken into account for the finalization of regulations and guides.

V-4.2.2. Challenges faced and lessons learned

The FANR stakeholders map has expanded since the establishment of the nuclear programme. The early communication plans for future nuclear projects and the readiness of the internal and external communication tools with different levels of stakeholders were the main success factors. Achieving the transparency and stakeholder engagement require specific tools enabling the sharing of information in a timely and secure manner. For example, the UAE nuclear programme involves experts around the world working on reviewing applications documents. A secured system was put in place to share information in very limited privilege for the users.

The IRRS mission to the UAE in 2011 was satisfied with the way communication and consultation with interested parties are established and implemented in FANR.

V-4.3. Leadership and management for safety

V-4.3.1. Implementation

In the UAE Nuclear Policy, the UAE pledged to work directly with the IAEA and to conform to its high standards in establishing a peaceful nuclear power programme. Those standards call upon nuclear organizations to promote and maintain what the IAEA calls a rigorous safety culture.

FANR has established through FANR Regulation on Leadership and Management for Safety in Nuclear Facilities (FANR-REG-01, V1) the requirements applying to the operating organization as regards leadership for safety and management for safety [V–35]. Such requirements are aligned with IAEA Safety Standards Series No. GSR Part 2, Leadership and Management for Safety [V–36].

Pursuant to the commitments set forth in the UAE Nuclear Policy and in line with the Convention on Nuclear Safety [V–12], and other relevant international instruments to which the UAE is a party, FANR has committed itself to core values which include safety culture as one of its core values in addition to transparency, collaboration, independence and excellence.

As regards the management system, FANR is also committed to adopting several measures in compliance with IAEA standards. FANR operates in line with its IMS, which was established according to GSR Part 2 [V–36]. The IMS integrates all requirements for safety, security and safeguards, it sets out the FANR mission, vision and core values, and incorporates structured policies, processes and procedures that have helped FANR to deliver its functions effectively and support the development of a strong safety culture.

The IMS includes a set of interacting processes that address the objectives and requirements of the organization. Elements included in the IMS that are tailored specifically to the regulator are the core processes addressing core regulatory functions, including the development and revision of regulations and guides, licensing, inspection and enforcement. These core processes are complemented by the management process (corporate management, etc) and support process (administrative functions).

The IMS Committee, which is chaired on a rotational basis by one of the two deputy directors general, is composed of all FANR directors. It meets regularly to openly discuss the implementation of all FANR processes. While the IMS process for self-evaluation and performance improvement empowers each FANR employee to report problems and identify opportunities for improvement.

In 2013, the IMS was incorporated into an electronic document management system as a means of integrating all FANR documents into a single document control programme.

All approved revisions of policies, processes and procedures, the IMS self-assessment, minutes of meetings, non-conformance reports and terms of reference are included in the electronic document management system.

FIG. V.1: Integrated management system process map

V-4.3.2.   Challenges faced and lessons learned

An IRRS mission to the UAE in 2011 confirmed that FANR has at an early stage, developed an integrated management system providing an important support function for the activities of the Authority.

In transition to operation, a project entitled 'Architecture of Integrated Information Systems' (ARIS) was initiated in 2019 as an enterprise tool used to have a holistic view of process/procedure modelling and analysis.

The ARIS objectives are:

- To improve and clarify the processes via a modern and easy-to-use user interface.
- Ensure consistency of processes and procedures and reuse of models and data through a central repository.

The FANR core processes are the backbone of FANR regulatory activities and oversight. In transition to operation, FANR introduced a regulatory oversight management system, the planning of which was initiated in 2018. This was established to respond to the need for an integrated system that includes all the licensee requests, reports, inspections with automated features that would support FANR in the conduct the regulatory oversight activities of the operation of the BNPP, and to measure the licensee performance during operation.

The objective was to deliver a system that would support FANR to:

- Prepare and issue inspection reports.
- Manage and close inspection findings.
- Capture, assess and close reportable events.
- Assess licensee change requests and approve or reject the request.
- Report plant status
- Capture, assess and communicate international operating experience and feedback (OPEX)
- Prepare, manage and control FANR inspection plans and notifications.
- Manage and control the Reactor Operator (RO)/Senior Reactor Operator (SRO) certification records.

V-4.4. Development of human resources

V-4.4.1. *Implementation*

The Government of the UAE has recognized the importance of human resources in the UAE Nuclear Policy. The UAE Nuclear Policy sets the basis for establishing a strategy to strengthen the human resources to regulate, manage, operate and maintain the safety of nuclear facilities.

The human resources strategy employed in the UAE nuclear power programme comprises two tracks:

- Staffing by senior experts including international staff to address the immediate and mid-term needs.
- Development of national capacity to ensure long-term sustainability.

Sufficient numbers of staff have been hired within FANR and ENEC and now Nawah to meet their business needs. These organizations continue to recruit qualified national and international experts.

A national capacity building effort have been implemented by ENEC, FANR and Khalifa University. The three entities are working together across education, training, and recruitment to ensure the nuclear programme's human resource needs are met in the longer term.

FANR human resource development

During Phase 1 of nuclear programme in the UAE, before the issuance of the Nuclear Law, the NEPIO played a role in the development of the emergent regulatory body. NEPIO recruited several experienced senior staff and advisors to assist in planning and development of the regulatory organization. At the national level, a base of technical expertise was brought in from other sectors as fundamental support for the programme.

Following the issuance of the Nuclear Law in 2009 [V–1], FANR was established as a regulatory body with legal, administrative, and financial independence. FANR management structure comprises at the top a Board of Management, the members of which are appointed by a resolution of the UAE Cabinet and constituted entirely by qualified nationals of the UAE. The Board of Management provides leadership and direction to FANR. The daily business of FANR is exercised under the direction of the Director General of FANR.

Significant efforts by FANR have been made in recruiting a core team of national and international experienced staff to develop the regulatory framework. FANR is a multi-cultural organization with more than 28 different nationalities distributed across the various departments. Everyone in the organization possesses knowledge that is integral to day-to-day work.

Measures to develop and maintain competence.

Having successfully recruited a workforce to meet near-term demands, FANR's human resource strategy for long-term sustainability concentrated on developing Emiratis to take positions of increasing responsibility while retaining an appropriate cadre of non-Emirati experts. FANR complemented its in-house training programmes through collaboration with ENEC, Khalifa University, the IAEA, and other partner institutions in a national programme of capacity-building, which offers Emiratis a range of education, training and development opportunities in the UAE and overseas [V–9].

Scholarships

In 2009, ENEC, FANR, and the Khalifa University (KU) for science, technology and research launched the UAE Nuclear Energy Scholarship Programme, which provides UAE nationals with a full scholarship to enrol in universities around the world to pursue a bachelor's or a master's degree in nuclear, mechanical, or electrical engineering. Once the studies are completed, the scholars have career opportunities in the UAE's growing nuclear energy industry.

Employee Development Programme

The Employee Development Programme is designed to support the development of FANR's employees by equipping them with the knowledge and skills needed to perform their roles and responsibilities.

The development of FANR existing and future managers was also a high priority for FANR. A specific development programme was implemented that includes management and leadership courses conducted in the UAE and abroad. This programme was included as a part of FANR's capacity-building approach to support Emiratis to take on leadership roles within FANR.

All FANR employees were offered to attend numerous in-house trainings and external courses covering technical skills, personal skills, and management and leadership topics. In-house training courses are coordinated by the Education and Training Department and delivered by staff experts and external consultants.

FANR employees were enrolled in Gulf Nuclear Energy Infrastructure Institute (GNEII), which is housed at Khalifa University (KU) for science, technology and research in partnership with Sandia National Laboratories and Texas A&M University.

GNEII provided an institutional capability for nuclear energy infrastructure development emphasizing education in nuclear energy safety, safeguards, and security. GNEII's goal is to be an indigenous, self-supporting regional education centre for future nuclear energy professionals working in government and industry. GNEII gave an opportunity for graduating fellows who work at ENEC, FANR and other industry-related organizations to share their research results and network with experts from around the world.

In addition, Khalifa University hosted the Nuclear Energy Management (NEM) School in 2012, 2015, and in 2017, which provided training courses delivered by the IAEA and organized in cooperation with FANR, for students and young professionals from Asia and the Pacific region. The courses provided relevant knowledge to ensure a solid nuclear expert foundation, through building leadership skills to manage nuclear power programmes.

A four-year cooperation agreement signed in October 2017 by the UAE and the IAEA designated the Khalifa University of Science and Technology as an IAEA Collaborating Centre. The objective of the collaborating centre is to share UAE's experience and expertise in nuclear infrastructure development, particularly through implementing specific training courses in collaboration with IAEA, and dispatching experts to support IAEA activities. The main activities of the collaborating centre included train professionals recommended by the IAEA in the field of nuclear power infrastructure development, develop and implement tailor-made training courses, workshops and work-plans for fellowship programmes, with the IAEA, to address specific issues relevant to embarking countries.

Knowledge management

FANR has established a knowledge management programme for business sustainability purposes. This programme has been designed to support its knowledge-based decisions for the safe and efficient regulation of nuclear activities in the UAE.

The knowledge management programme aims to support FANR management by minimizing the impact of employee mobility (e.g. transfer of personnel within or outside the organization, retirement), and preserving knowledge. It also facilitated the transfer of nuclear knowledge from one employee to the next.

V-4.4.2. Challenges faced and lessons learned

The factors that supported the rapid development of the human resources included clear government policy, leadership and detailed planning, combined with the recruitment of experienced national and international staff to launch the regulatory programme.

The active coordination in human resources development undertaken in the different organizations, especially between the utility, the regulatory body, and the educational community, is a model for the effective use of resources. It also helped ensure an integrated approach to the development of required workforce and competence.

An IRRS mission to the UAE in 2011 confirmed that students who participated in capacity-building programmes had the opportunity to choose at the commencement of their career, to work for either the regulator FANR or for the operator ENEC. This is considered as a good contribution towards the balanced development of the human capacity throughout the whole nuclear sector.

More recently, FANR hosted in November 2023 an expert mission dispatched by the IAEA to evaluate the regulatory capacity building system. The expert mission identified as a good practice that FANR has developed well-designed processes and related documents regarding the four pillars of capacity building to ensure a systematic and consistent approach is followed in the development and maintenance of human capacity. These pillars are education and training, human resources management, knowledge management and knowledge networks.

REFERENCES TO ANNEX V

[V–1]   INTERNATIONAL ATOMIC ENERGY AGENCY, Integrated Regulatory Review Service (IRRS) Mission to the United Arab Emirates (2011). https://www.iaea.org/sites/default/files/documents/review missions/irrs_mission_to_uae_dec_2011.pdf

[V–2]   UNITED ARAB EMERITES GOVERNMENT, Policy of the United Arab Emirates on the Evaluation and Potential Development of Peaceful Nuclear Energy", April 2008.

[V–3]   INTERNATIONAL ATOMIC ENERGY AGENCY, Mission Report on the Integrated Nuclear Infrastructure Review (INIR) – PHASE 3 (2018). https://www.iaea.org/sites/default/files/documents/review-missions/2018-uae-inir-phase-3-report.pdf

[V–4]   INTERNATIONAL ATOMIC ENERGY AGENCY, Milestones in the Development of a National Infrastructure for Nuclear Power, IAEA Nuclear Energy Series No. NG-G-3.1 (Rev. 2), IAEA, Vienna (2024), https://doi.org/10.61092/iaea.zjau-e8cs

[V–5]   INTERNATIONAL ATOMIC ENERGY AGENCY, Report on the Integrated Nuclear Infrastructure Review (INIR) Mission to Review the Status of the National Nuclear Infrastructure in The United Arab Emirates (2011). https://www.iaea.org/sites/default/files/documents/review-missions/uae-inir-mission-january-2011-report.pdf

[V–6]   EUROPEAN ATOMIC ENERGY COMMUNITY, FOOD AND AGRICULTURE ORGANIZATION OF THE UNITED NATIONS, INTERNATIONAL ATOMIC ENERGY AGENCY, INTERNATIONAL LABOUR ORGANIZATION, INTERNATIONAL MARITIME ORGANIZATION, OECD NUCLEAR ENERGY AGENCY, PAN AMERICAN HEALTH ORGANIZATION, UNITED NATIONS ENVIRONMENT PROGRAMME, WORLD HEALTH ORGANIZATION, Fundamental Safety Principles, IAEA Safety Standards Series No. SF-1, IAEA, Vienna (2006), https://doi.org/10.61092/iaea.hmxn-vw0a

[V–7]   INTERNATIONAL ATOMIC ENERGY AGENCY, Governmental, Legal and Regulatory Framework for Safety, IAEA Safety Standards Series No. GSR Part 1 (Rev. 1), IAEA, Vienna (2016).

[V–8]   Vienna Convention on Civil Liability for Nuclear Damage, INFCIRC/500, IAEA, Vienna (1963).

[V–9]   UNITED ARAB EMIRATES, UAE national report for the Joint 8th and 9th Review Meetings of the Convention on Nuclear Safety (2023). https://www.iaea.org/sites/default/files/24/02/cns_uae_national_report_2022.pdf

[V–10] UNITED ARAB EMIRATES, UAE national report for the Seventh Review Meeting of the Convention on Nuclear Safety (2017). https://www.iaea.org/sites/default/files/uae_nr-7th-rm.pdf.

[V–11] Joint Protocol Relating to the Application of the Vienna Convention and the Paris Convention INFCIRC/402, IAEA, Vienna (1988).

[V–12] Convention on Nuclear Safety, INFCIRC/449, IAEA, Vienna (1994).

[V–13] Joint Convention on the Safety of Spent Fuel Management and on the Safety of Radioactive Waste Management, INFCIRC/546, IAEA, Vienna (1997).

[V–14] FEDERAL AUTHORITY FOR NUCLEAR REGULATION, Fourth National Report of the United Arab Emirates, 6th Review Meeting of the Joint Convention on Safety of Spent Fuel Management and Safety of Radioactive Waste Management, Vienna, May 2018, Power Point Presentation, FANR (2020)

[V–15] Convention on Early Notification of a Nuclear Accident, INFCIRC/335, IAEA, Vienna (1986).

[V–16] Convention on Assistance in the Case of a Nuclear Accident or Radiological Emergency, INFCIRC/336, IAEA, Vienna (1986).

[V–17] Convention on the Physical Protection of Nuclear Material, INFCIRC/274/Rev. 1, IAEA, Vienna (1979).

[V–18] Amendment to the Convention on the Physical Protection of Nuclear Material, INFCIRC/274/Rev. 1, IAEA, Vienna (2005).

[V–19] Treaty on the Non-Proliferation of Nuclear Weapons, INFCIRC/140, IAEA, Vienna (1986).

[V–20] Additional Protocol to CSA, INFCIRC/540, IAEA, Vienna (1997).

[V–21] Protocol to Amend the Vienna Convention on Civil Liability for Nuclear Damage, INFCIRC/566, IAEA, Vienna (1997).

[V–22] Joint Protocol Relating to the Application of the Vienna Convention and the Paris Convention INFCIRC/402, IAEA, Vienna (1988).

[V–23] Revised Supplementary Agreement Concerning the Provision of Technical Assistance by the IAEA, INFCIRC/267, IAEA, Vienna (1997).

[V–24] INTERNATIONAL ATOMIC ENERGY AGENCY, Code of Conduct on the Safety and Security of Radioactive Sources and its supplementary Guidance on the Import and Export of Radioactive Sources, IAEA, Vienna (2004)

[V–25] INTERNATIONAL ATOMIC ENERGY AGENCY, Communication dated 20 January 2022 received from the Permanent Mission of the United Arab Emirates to the Agency, INFCIRC/969, IAEA, Vienna (2022)

[V–26] INTERNATIONAL ATOMIC ENERGY AGENCY, Integrated Regulatory Review Service (IRRS) Extended Follow-Up Mission to United Arab Emirates (2015). https://www.iaea.org/sites/default/files/documents/review-missions/uae_irrs_follow-up_mission_report.pdf

[V–27] INTERNATIONAL ATOMIC ENERGY AGENCY, Peer Review of the Arrangements in the United Arab Emirates Regarding the Preparedness for Responding to a Nuclear Emergency at the Barakah Nuclear Power Plant (2015).

[V–28] INTERNATIONAL ATOMIC ENERGY AGENCY, IAEA Sees Safety Commitment at UAE's First Nuclear Power Plant Ahead of Planned Operation Start (2017). https://www.iaea.org/newscenter/pressreleases/iaea-sees-safety-commitment-at-uaes-first-nuclear-power-plant-ahead-of-planned-operation-start

[V–29] Country Nuclear Power Profiles | IAEA (2021). https://www-pub.iaea.org/MTCD/Publications/PDF/CNPP2021/countryprofiles/UnitedArabEmirates/UnitedArabEmirates.html.

[V–30] FEDERAL AUTHORITY FOR NUCLEAR REGULATION, Regulation. Siting of Nuclear Facilities (FANR-REG-02), FANR, Abu Dhabi (2013).

[V–31] FEDERAL AUTHORITY FOR NUCLEAR REGULATION, Regulation Design of Nuclear Power Plants (FANR-REG-03), FANR, Abu Dhabi (2013).

[V–32] FEDERAL AUTHORITY FOR NUCLEAR REGULATION, Regulation Application for a Licence to Construct a Nuclear Facility, (FANR-REG-06), FANR, Abu Dhabi (2010).

[V–33] FEDERAL AUTHORITY FOR NUCLEAR REGULATION, Regulation for an Application for a License to Operate a Nuclear Facility, (FANR-REG-14), FANR, Abu Dhabi (2010).

[V–34] FEDERAL AUTHORITY FOR NUCLEAR REGULATION, Operational Safety including Commissioning, (FANR-REG-16), FANR, Abu Dhabi (2010).

[V–35] FEDERAL AUTHORITY FOR NUCLEAR REGULATION, Management Systems for Nuclear Facilities (FANR-REG-1), FANR, Abu Dhabi (2011).

[V–36] INTERNATIONAL ATOMIC ENERGY AGENCY, Leadership and Management for Safety, IAEA Safety Standards Series No. GSR Part 2, IAEA, Vienna (2016).

VI-1. EXISTING SAFETY INFRASTRUCTURE PRIOR TO INITIATING A NUCLEAR POWER PROGRAMME IN UZBEKISTAN

The State inspectorate on Supervision of Geological Study of Mineral Resources and Safety in Industry, Mining and the Public Sector (Sanoatgeokontekhnazorat) is the main regulatory body in the field of ensuring industrial, geological, mining and radiation and nuclear safety. The structure of the State inspectorate consists of several departments including the department on atomic supervision. In the department on atomic supervision there have only been six inspectors working. The main two nuclear facilities are the WWR-SM pool-type nuclear research reactor and the Photon impulse type nuclear research reactor. In addition, all industrial and medical facilities that use ionizing radiation sources are also regulated by the State inspectorate Sanoatgeokontekhnazorat.

VI-2. STATUS OF NUCLEAR POWER PROGRAMME IN UZBEKISTAN

In September 2018, an intergovernmental agreement was signed between Uzbekistan and the Russian Federation on the construction by Rosatom of two VVER-1200 reactors.

On July 19, 2018, Decree of the President of the Republic of Uzbekistan No. 5484 "On measures for the development of nuclear energy in the Republic of Uzbekistan" was approved [VI–1].

A priority site for the construction of the nuclear power plant (NPP) has been selected and a survey of the site has been carried out. From 16 to 20 January 2023, the IAEA conducted a Site Design and External Events (SEED) review service mission [VI–2].

At present preparatory work is under way to obtain permits for the site.

VI-3. ROLES AND RESPONSIBILITIES OF THE GOVERNMENT AND THE REGULATORY BODY IN UZBEKISTAN IN PHASE 1 OF THE NUCLEAR POWER PROGRAMME

VI-3.1. National policy and strategy for safety

VI-3.1.1.   Implementation

On July 19, 2018, by the Decree of the President of the Republic of Uzbekistan No. 5484 "On Measures for the Development of Nuclear Energy in the Republic of Uzbekistan" [VI–1], a nuclear energy programme implementing organization (NEPIO) authorized for the development and implementation of a unified State policy and strategic direction in the field of nuclear energy development was established - the Agency for the Development of Nuclear Energy (Uzatom).

By Decree of the President of the Republic of Uzbekistan dated December 12 2018, No. 5594 "On measures to cardinally improve the system of state administration and supervision in the spheres of industrial, radiation and nuclear safety", the State Committee on Industrial Safety (SCIS) was established [VI–3]. The SCIS is an authorized government body responsible for implementing a unified State policy and exercising control in the field of ensuring radiation and nuclear safety at nuclear power and nuclear technology facilities, as well as in the field of industrial safety at hazardous production facilities.

On December 12 2018 the Resolution of the President of the Republic of Uzbekistan, No. 4058 "On the organization of the activity of the State Committee on Industrial Safety of the Republic of Uzbekistan" was approved [VI–4].

In accordance with the Resolution of the President of the Republic of Uzbekistan dated February 7, 2019, No. PP-4165, the Concept for the Development of Nuclear Energy in the Republic of Uzbekistan for 2019-2029 and the road map for its implementation were approved. The State Programme for the Development of Nuclear Energy in Uzbekistan was developed and submitted for approval in accordance with the developed concept.

VI-3.1.2. *Challenges faced and lessons learned*

In the process of developing and approving a national policy and strategy, IAEA safety standards and countries with a similar legislative system were considered.

VI-3.2. Governmental and legal framework for safety

VI-3.2.1. *Implementation*

In order to create a regulatory framework for the use of atomic energy, the draft Law of the Republic of Uzbekistan "On the Use of Atomic Energy for Peaceful Purposes" was developed and approved in the second and third readings by the Legislative Chamber of the Oliy Majlis (parliament of Uzbekistan) in 2019 [VI–5].

In 2019, the law "On the Use of Atomic Energy for Peaceful Purposes" was adopted [VI–6], and legislation in the field of the use of atomic energy is being improved. According to Article 16 of the atomic law, the specially authorized body for State regulation of the safety of the use of atomic energy is the SCIS. In addition, the atomic law sets out the functions and responsibilities of the regulatory body.

In addition, the Resolution of the Cabinet Ministers of the Republic of Uzbekistan No. 75 "On approval of provision of State Committee on Industrial Safety of the Republic of Uzbekistan" which defines the functions and responsibilities of the regulatory body was approved in 2019 [VI–5].

The Concept for the Development of Nuclear Energy in the Republic of Uzbekistan for the period 2019-2029 and the Strategy for the Development of Human Resources for the country's nuclear energy programme were adopted and SCIS was defined as one of the responsible organizations for performing tasks in the framework of the Concept [VI–5].

VI-3.2.2. *Challenges faced and lessons learned*

With the creation of the new atomic law, in which the roles and responsibilities of the main regulatory body and other regulatory bodies are clearly defined, as well as the NEPIO's, the areas of use of atomic energy were also defined. During the process of developing the atomic law, the experience of leading countries' legislative frameworks was studied, and the IAEA safety standards were also considered. Before development of the atomic law, the existing legislative framework of the country was reviewed and assessed.

VI-3.3.1.   Implementation

Under the 2019 Atomic Law [VI–6], SCIS was appointed as the regulatory body for State regulation responsible for the safe use of atomic energy as well as for developing, approving, and enforcing the application of regulations and rules in the field of atomic energy use [VI–6].

In accordance with the atomic law there are other regulatory bodies such as the State Committee on Ecology and Environmental Protection, Ministry of Emergency Situations, Ministry of Health, Ministry of Construction, the State Committee on Geology and Mineral Resources, the State Customs Committee, the Centre of Hydrometeorological Service of the Cabinet of Ministers (Uzhydromet), the Ministry of Internal Affairs, the State Security Service, the Ministry of Defence, and the National Guard. Final decisions are made by SCIS which is the main regulatory body and coordinates with other regulatory bodies.

Under Presidential Resolution No. 4058, the Department on Radiation and Nuclear Safety was established as part of SCIS. The organizational structure of the Department includes 5 units with 14 divisions. SCIS is mandated to coordinate its activities with the other State bodies with regulatory responsibilities related to safety, security and safeguards [VI-2].

As of March 2023, several regulations have been developed such as regulations for licensing a new nuclear power plant (NPP), issuing site permits and conducting safety assessments, as well as procedures for preparing, training, and retraining nuclear installation personnel, and procedures for the conduct and organization of inspections.

According to the decree of the President of the Republic of Uzbekistan No. PD-5594 the decisions of the SCIS adopted within its powers, are binding on State and economic management bodies [VI–3], local executive authorities, as well as other organizations and individuals.

The Chair and deputy Chair of SCIS are appointed by the President of the Republic of Uzbekistan on the proposal of the Prime Minister of the Republic of Uzbekistan. The Resolution of the Cabinet Ministers on "Approving of provision of SCIS" defines the regulatory body's responsibilities with regard to quality, results and consequences of the implementation of draft regulations and other documents submitted to the Cabinet of Ministers and the Administration of the President of the Republic of Uzbekistan, as well as decisions made by SCIS [VI–5].

Oversight activities (review, assessment and inspections) will be conducted in accordance with national regulations.

Norms and standards of the vendor country have been adopted for review and assessment of the NPP designs. Review and assessment of the application for the site permit and licence applications and inspection during construction is planned to be supported by local and/or foreign technical support organizations.

VI-3.3.2.   Challenges faced and lessons learned

The creation of an independent regulatory body empowered the regulatory body to make its own decisions on issuing or rejecting licences and permits. In addition, the regulatory body can develop regulations and guides on ensuring the safety and security of facilities' use of atomic energy.

According to the Presidential decree on establishing the State Committee on Industrial Safety of Uzbekistan, the Department of Radiation and Nuclear Safety was created as a part of SCIS without the formation of a legal entity [VI–5]. Specifically, the Department of Radiation and

Nuclear Safety plan for human resources was to engage up to 71 full-time personnel, funded from the State Budget of Uzbekistan.

The COVID-19 pandemic affected the economic state of all countries in the world and consequently constrained the full implementation of the regulatory body in Uzbekistan.

There were also constraints in inviting an IRRS mission due to the lack of technical specialists to complete the self-assessment. As a result, the IRRS mission has been postponed two times.

However, a Phase 2 INIR mission was conducted in 2021 [VI–5]. .

VI-3.4. Other governmental considerations – global nuclear safety regime

VI-3.4.1.   Implementation

On October 8 1994, an Agreement was signed between the Republic of Uzbekistan and the International Atomic Energy Agency on the application of safeguards in connection with the Treaty on the Non-Proliferation of Nuclear Weapons [VI–7] and on September 22, 1998, an Additional Protocol to it [VI–8]. This Agreement was approved by the Resolution of the Cabinet of Ministers of the Republic of Uzbekistan dated May 6, 2004, No. 212 "On Approval of an International Treaty". In addition, a resolution of the Cabinet of Ministers dated June 25, 2009, No. 179 "On measures to fulfil the obligations of the Republic of Uzbekistan under international treaties in the field of peaceful use of atomic energy" was adopted [VI–5] [VI–9].

By the Decree of the Oliy Majlis of the Republic of Uzbekistan dated December 26, 1997, No. 556-I, the Republic of Uzbekistan acceded to the Convention on the Physical Protection of Nuclear Material [VI–9].

By Resolution of the President of the Republic of Uzbekistan dated January 10, 2013, No. 1906, Uzbekistan acceded to the amendment to the Convention on the Physical Protection of Nuclear Material [VI–9] [VI-10].

By the Law of the Republic of Uzbekistan dated December 11, 2008, No. ZRU-186, Uzbekistan acceded to the Joint Convention on the Safety of Spent Fuel Management and on the Safety of Radioactive Waste Management [11] .

In addition, Uzbekistan joined the Treaty on a Nuclear-Weapon-Free Zone in Central Asia (Semipalatinsk, September 8, 2006) [VI–10]. At 2006 Uzbekistan ratified by the Law of the Republic of Uzbekistan dated April 2 2007 "On ratification of the Treaty on a Nuclear-Weapon-Free Zone in Central Asia (Semipalatinsk, September 8, 2006)" [VI–13].

A memorandum of understanding was signed in 2018 between SCIS and Rostekhnadzor on cooperation in the field of regulation of industrial, nuclear and radiation safety in the field of the use of atomic energy for peaceful purposes.

In 2021, an agreement was signed between SCIS and the Ministry of Emergency Situations of Belarus on cooperation in ensuring industrial, nuclear and radiation safety in the use of atomic energy.

Also in 2021, an agreement was signed between the government of the Russian Federation and the government of Uzbekistan on prompt notification of a nuclear accident and exchange of information in the field of nuclear and radiation safety.

Another agreement was signed in 2021 between the SCIS and Rostekhnadzor on cooperation in the field of regulation of nuclear and radiation safety in the field of atomic energy use for peaceful purposes.

In 2022, a Memorandum of Understanding was established between SCIS and the Norwegian Radiation and Nuclear Safety Authority on cooperation in the field of nuclear and radiation safety.

Also in 2022, a Memorandum of Understanding was established between the Atomic Energy Regulatory Board of India and SCIS for the exchange of information and cooperation in the regulation of radiation and nuclear safety in the use of atomic energy for peaceful purposes.

As of 2023, a Memorandum of Understanding was under development between the SCIS and the Nuclear Safety and Security Commission of the Republic of Korea for cooperation in the regulation of radiation and nuclear safety in the field of atomic energy for peaceful purposes.

VI-3.4.2. Challenges faced and lessons learned

The Road map for the implementation and the development of nuclear energy in Uzbekistan for the period 2019-2029 was approved by the Decree of the President of the Republic of Uzbekistan dated February 7, 2019, No. PP-4165 [VI–14]

GP: establishment of 10-year plan 2019-2029 for the Republic of Uzbekistan to join the following international conventions:

- Vienna Convention on Civil Liability for Nuclear Damage[VI–15].
- Convention on Early Notification of a Nuclear Accident [VI–16].
- Convention on Nuclear Safety [VI–17].
- Convention on Assistance in the Case of a Nuclear Accident or Radiological Emergency [VI–18].

Note: All of them are undergoing Governmental approval

VI-4. KEY PRIORITIES FOR THE GOVERNMENT AND THE REGULATORY BODY IN UZBEKISTAN IN PHASE 1

VI-4.1. Site survey

VI-4.1.1. Implementation

Requirements are being prepared in advance to prepare for regulatory review.

According to the 2019 Atomic Energy Act [VI–6], SCIS is responsible for issuing site permits for nuclear facilities after review and assessment of the site evaluation report. The procedure for review and assessment is defined in the Resolution of the Cabinet of Ministers No. 390 of June 17, 2020. In addition, a Resolution Cabinet of Ministers of the Republic of Uzbekistan on approval of the regulations for issuing a permit for the use of a site for the placement of a nuclear installation and/or a nuclear waste storage facility was approved in 2022 [VI–19].

A Resolution of the Cabinet of Ministers of the Republic of Uzbekistan "On the approval of the regulation on the procedure for issuing a permit for the use of a site for placing a nuclear device and (or) a storage facility" was approved on March 28, 2022.

An internal document of the regulatory body on requirements for placement of storage facilities for nuclear materials and radioactive substances is under development [VI–5].

VI-4.1.2.   Challenges faced and lessons learned

Several documents have been developed and approved regarding site permitting. However, the staff of the regulatory body has no experience on how to develop internal documents on procedures for reviewing documents submitted by the applicant.

There needs to be cooperation with other countries' regulatory bodies in the early stages of the project to ensure competent staff for the issuing of site permits. In addition, it is necessary to participate in the IAEA's international training courses and workshops on reviewing documents submitted for site permit applications.

VI-4.2. Consultation with interested parties

VI-4.2.1.   Implementation

On March 3 2020, an Action Plan was approved for holding public consultations and disclosure of information, as well as involving the public in the environmental impact assessment of the construction and operation of nuclear power plants in Uzbekistan [VI–5], in accordance with which measures are provided for organizing public hearings on the territory of Uzbekistan and neighbouring countries.

In addition, Resolution of the President of the Republic of Uzbekistan dated February 7, 2019, No. PP-4165 "On approval of the concept for the development of nuclear energy in the Republic of Uzbekistan for the period 2019-2029" provides for ensuring transparency and openness to the public regarding the nuclear energy programme [VI–14].

According to the Resolution of the Cabinet of Ministers of the Republic of Uzbekistan No. 75 "On approval of the Regulations on the State Committee on Industrial Safety of the Republic of Uzbekistan" was defined that, among other things, reflects the strategy of the regulatory body regarding the availability of information to the public, as well as regulatory communication and consultations with stakeholders [VI–5].

VI-4.3. Leadership and management for safety

VI-4.3.1.   Implementation

The State Committee on Industrial Safety has started to establish it integrated management system (IMS) including safety culture. An internal working group has been created for the development of a road map for establishing the IMS. The process started with support of the IAEA on conduction of two fallowing national workshops:

- National Workshop on Establishment of Integrated Management System in the Regulatory Body.
- National Workshop on Nuclear Security Culture in Practice.

VI-4.3.2.   Challenges faced and lessons learned

The IMS takes a long time to develop, so needs to be done in the early stages of establishment of the regulatory body.

VI-4.4. Development of human resources

VI-4.4.1. Implementation

In order to provide human resources for the implementation of the national nuclear energy programme, the Resolution of the President of the Republic of Uzbekistan dated October 16, 2019, No. PP-4492 "On approval of the Strategy for the development of human resources for the nuclear energy programme of the Republic of Uzbekistan" was adopted [VI–20].
According to the above Presidential Resolution, in the period 2019-2030, continuous training of personnel will be provided for all organizations involved in the implementation of the nuclear energy programme and measures will be taken to retain industry specialists.
In addition, the Resolution of the President of the Republic of Uzbekistan dated December 12, 2018, No. PP-4058 "On the organization of the activities of the State Committee for Industrial Safety of the Republic of Uzbekistan" approved the staffing schedule for the Department of Radiation and Nuclear Safety in the period 2018-2028 [VI–4].
SCIS is further developing its competence through a programme of cooperation with Rostekhnadzor. It has concluded following cooperation agreements and memorandums of understanding [VI–5]:

- Agreement between the government of the Russian Federation and the government of Uzbekistan on prompt notification of a nuclear accident and exchange of information in the field of nuclear and radiation safety
- Agreement between the State Committee on Industrial Safety of Uzbekistan (SCIS) and the Ministry of Emergency Situations of Belarus on cooperation in ensuring industrial, nuclear and radiation safety in the use of atomic energy.
- Agreement between SCIS and Rostekhnadzor on cooperation in the field of regulation of nuclear and radiation safety in the field of atomic energy use for peaceful purposes.
- Memorandum of understanding between SCIS and Rostekhnadzor on cooperation in the field of regulation of industrial, nuclear and radiation safety in the field of atomic energy use for peaceful purposes.
- Memorandum of understanding between SCIS and Norwegian Radiation and Nuclear Safety Authority on cooperation in the field of nuclear and radiation safety.
- Memorandum of understanding between the Atomic Energy Regulatory Board of India and SCIS for the exchange of information and cooperation in the regulation of radiation and nuclear safety in using atomic energy for peaceful purposes.
- Further agreements and memorandums of understanding with relevant organizations of the Republic of Korea, Türkiye and China are under development.
-

In addition, SCIS cooperates with the IAEA in the framework of the technical cooperation programme on a national project on Strengthening the Regulatory Framework and Infrastructure for Effective Regulation and Regulatory Oversight of NPPs, Radiation Facilities and Sources.

VI-4.4.2. Challenges faced and lessons learned

As per the existing staffing plan, the number of staff to be recruited by 2020 was envisaged to be 30 while the number currently stands at 18. Budgetary constraints had a major impact on the level of salaries and the ability to fill vacancies. There is also a problem with the training of personnel and the lack of modern tools for conducting inspections [VI-2].

Uzbekistan can learn from the experience of mature regulatory bodies on the establishment of knowledge management, as well as taking into consideration the IAEA safety standards.

SCIS is empowered to engage internal and external technical support organizations to support licensing and regulatory oversight.

A resolution of Cabinet of Ministers on enhancing capabilities of regulatory body was drafted and is currently pending approval. Once approved, financial support and human resource development issues will be resolved.

REFERENCES TO ANNEX VI

[VI–1] REPUBLIC OF UZBEKISTAN, Decree of the President of the Republic of Uzbekistan No. 5484 on Measures for the Development of Nuclear Energy in the Republic of Uzbekistan, 19 July 2018, Tashkent (2018).

[VI–2] INTERNATIONAL ATOMIC ENERGY AGENCY, IAEA Team in Uzbekistan Concludes Site and External Events Design (SEED) Review for the Country's First Nuclear Power Plant (2023). https://www.iaea.org/newscenter/pressreleases/iaea-team-in-uzbekistan-concludes-site-and-external-events-design-seed-review-for-the-countrys-first-nuclear-power-plant

[VI–3] REPUBLIC OF UZBEKISTAN, Decree of the President of the Republic of Uzbekistan No. DP-5594 on Measures to Radically Improve the System of State Management and Supervision in the Fields of Industrial, Radiation and Nuclear Safety, 12 December 2018, Tashkent (2018).

[VI–4] REPUBLIC OF UZBEKISTAN, Resolution of the President of the Republic of Uzbekistan No.RP-4058 on the Organization of Activities of the State Committee on Industrial Safety of the Republic of Uzbekistan, 12 December 2018, Tashkent (2018)

[VI–5] INTERNATIONAL ATOMIC ENERGY AGENCY, Mission Report on the Integrated Nuclear Infrastructure Review Mission Phase 2 to Uzbekistan, 24 May to 3 June 2021. https://www.iaea.org/sites/default/files/documents/review-missions/inir2-uzbekistan-030621.pdf

[VI–6] REPUBLIC OF UZBEKISTAN, Law of the Republic of Uzbekistan No. ZRU-565 on the Use of Atomic Energy for Peaceful Purposes, 9 September 2019, Tashkent (2019).

[VI–7] Treaty on the Non-Proliferation of Nuclear Weapons, INFCIRC/140, IAEA, Vienna (1986).

[VI–8] Additional Protocol to CSA, INFCIRC/540, IAEA, Vienna (1997).

[VI–9] The Convention on the Physical Protection of Nuclear Material, INFCIRC/274/Rev. 1, IAEA, Vienna (1980).

[VI–10] Amendment to the Convention on the Physical Protection of Nuclear Material, INFCIRC/274/Rev. 1/Mod. 1, IAEA, Vienna (2005).

[VI–11] Joint Convention on the Safety of Spent Fuel Management and on the Safety of Radioactive Waste Management, INFCIRC/546, IAEA, Vienna (1997).

[VI–12] UNITED NATIONS OFFICE FOR DISARMAMENT AFFAIRS, Central Asian Nuclear-Weapon-Free Zone Treaty, UNODA, Vienna (2006).

[VI–13] UNITED NATIONS OFFICE FOR DISARMAMENT AFFAIRS, Treaty on a Nuclear-Weapon-Free Zone in Central Asia | United Nations Platform for Nuclear-Weapon-Free Zones. https://www.un.org/nwfz/content/treaty-nuclear-weapon-free-zone-central-asia

[VI–14] REPUBLIC OF UZBEKISTAN, Decree of the President of the Republic of Uzbekistan, No. DP-4165 On Approval of the Concept for the Development of Nuclear Energy for the Period 2019-2029, 7 February 2019, Tashkent (2019).

[VI–15] Vienna Convention on Civil Liability for Nuclear Damage, INFCIRC/500, IAEA, Vienna (1963).

[VI–16] Convention on Early Notification of a Nuclear Accident, INFCIRC/335, IAEA, Vienna (1986).

[VI–17] Convention on Nuclear Safety, INFCIRC/449, IAEA, Vienna (1994).

[VI–18] Convention on Assistance in the Case of a Nuclear Accident or Radiological Emergency, INFCIRC/336, IAEA, Vienna (1986).

[VI–19] REPUBLIC OF UZBEKISTAN, Resolution of the Cabinet of Ministers of the Republic of Uzbekistan No. 390 on Approval of the Regulations on the Procedure for Examination of Safety Justification of Atomic Energy Facilities and (or) Activities in the Field of Atomic Energy Use, 17 June 2020, Tashkent (2020)

[VI–20] REPUBLIC OF UZBEKISTAN, Resolution the President of the Republic of Uzbekistan No. RP-4492 on Approving a Strategy for Developing Human Resources Potential for the Nuclear Energy Program of the Republic of Uzbekistan, 16 October 2019, Tashkent (2019).

LIST OF ABBREVIATIONS

AFRA	African Regional Cooperative Agreement for Research, Development and Training Related to Nuclear Science and Technology
ARIS	architecture of integrated information systems
AMSSNUR	Moroccan Agency for Nuclear and Radiological Safety and Security
BPTC	basic professional training course
CICPA	Infrastructure and Coastal Protection Authority (UAE)
CIS	Commonwealth of Independent States
CNIA	Critical National Infrastructure Authority (UAE)
CNPP	country's nuclear power profile
CNS	Convention for Nuclear Safety
CNSC	Canadian Nuclear Safety Commission
CSA	comprehensive safeguards agreement
EMS	electronic management system
ENEC	Emirates Nuclear Energy Company
ENNRA	Egyptian Nuclear and Radiological Regulatory Authority
EIA	Environmental Impact Assessment
ELSE	Tutoring and Training activities and European Leadership for Safety Education
ENSTTI	European Nuclear Safety Tutoring and Training Institute
EPC	engineering, procurement and construction
EPREV	Emergency Preparedness Review International
EU	European Union
ENSREG	European Nuclear Safety Regulators Group
EuCAS	European and Central Asia Safety Network
FANR	Federal Authority for Nuclear Regulation
FNRBA	Forum of Nuclear Regulatory Bodies in Africa
GNEII	Gulf Nuclear Energy Infrastructure Institute
GNPPO	Ghana Nuclear Power Programme
Gosatomnadzor	Department for Nuclear and Radiation Safety
Gospromatomnadzor	State Committee for Supervision of Industrial and Nuclear Safety
GRM	Generic Road Map
HEU	high enriched uranium
HR	human resources
HRD	human resource development
ICSANT	International Convention on Suppression on Acts of Nuclear Terrorism
IGA	inter-governmental agreement
IMS	integrated management system
INLEX	International Expert Group on Nuclear Liability
INS	International Nuclear Security of USDOE
INSC	International Nuclear Safety Cooperation
INSEP	International Nuclear Safeguards Engagement Programme
INIR	Integrated Nuclear Infrastructure Review
IPPAS	Physical Protection Advisory Service
IRDP	International Regulatory Development Partnership

IRRS	Integrated Regulatory Review Service
IRS	Incident Reporting Systems for Nuclear Installations
ISSAS	IAEA State system for accounting and control of nuclear material Advisory Service
KEPCO	Korea Electric Power Corporation
KINS	Korea Institute of Nuclear Safety
KU	Khalifa University of Science and Technology (UAE)
KSA	knowledge, skills, and attitudes
NSSC	Korea's Nuclear Safety and Security Commission
LEU	low enriched uranium
MESTI	Ministry of Environment Science Technology and Innovation (Ghana)
MOU	memorandum of understanding
NAEC	Nigeria Atomic Energy Commission
MOERE	Minister of Electricity and Renewable Energy (Egypt)
Nawah	Nawah Energy Company
NCNSRC	National Centre for Nuclear Safety and Radiation Control (Egypt)
NEPIC	Nuclear Energy Programme Implementation Committee (Nigeria)
NEPIO	nuclear energy programme implementing organization
NGOs	non-governmental organizations
NNRA	Nigerian Nuclear Regulatory Authority
NPG	Nuclear Power Ghana
NPP	nuclear power plant
NPPA	Nuclear Power Plants Authority (Egypt)
NPT	Non-Proliferation of Nuclear Weapons
NRA	Nuclear Regulatory Authority (Ghana)
NSDD	Nuclear Smuggling Detection and Deterrence of USDOE
OECD NEA	Organisation for Economic Co-operation and Development
ORS	Office of Radiological Security of USDOE
PCR	Programme Comprehensive Report
PJSC	Public Joint Stock Company (UAE)
PNRA	Pakistan Nuclear Regulatory Authority
PSAR	preliminary safety analysis report
pre-OSART	pre-Operational Safety Review Team
RAMP	Radiation Protection Computer Code Analysis and Maintenance Programme
RCF	Regulatory Cooperation Forum
RO	reactor operator
Rostekhnadzor	Federal Service for Environmental, Technological and Nuclear Supervision of the Russian Federation
SAC	Situation and Analytical Centre (Egypt)
SAR	safety analysis report
SARCON	Systematic Assessment of Regulatory Competence Needs
SCIS	State Committee on Industrial Safety of the Republic of Uzbekistan
SEED	Site and External Events Design
SER	site evaluation report
SNSA	School of Nuclear and Allied Sciences

SRO	senior reactor operator
SSC	systems, structures and components
STC NRS	Scientific and Technical Centre for Nuclear and Radiation Safety (Belarus)
TC	technical cooperation
TSO	technical support organization
UAE	United Arab Emirates
USNRC	United States Nuclear Regulatory Commission
Uzatom	Agency for the Development of Nuclear Energy (Uzbekistan)
VVER	water cooled, water moderated power reactor
WANO	World Association of Nuclear Operators

CONTRIBUTORS TO DRAFTING AND REVIEW

Abdulrauf, L.	Nuclear Regulatory Authority, Nigeria
Al Mansouri, N.	Federal Authority for Nuclear Regulation, United Arab Emirates
Al Mazrouei, M.	Federal Authority for Nuclear Regulation, United Arab Emirates
Ampomah-Amoako, E.	Nuclear Regulatory Authority, Ghana
Aoki, M.	International Atomic Energy Agency
Ashirmetov, A.	State Committee on Industrial Safety, Uzbekistan
Danilenka, N.	Gosatomnadzor, Belarus
Gaheen, M.	Nuclear and Radiological Regulatory Authority, Egypt
Gomaa, R.	International Atomic Energy Agency
Grant, I.M.	Consultant, Canada
Jubin, J.-R.	International Atomic Energy Agency
Khouaja, H.	International Atomic Energy Agency
Sobolev, O.	Gosatomnadzor, Belarus

Consultants' meetings

Vienna, Austria: 4–7 October 2022
Vienna, Austria: 14–17 March 2023

CONTACT IAEA PUBLISHING

Feedback on IAEA publications may be given via the on-line form available at:
www.iaea.org/publications/feedback

This form may also be used to report safety issues or environmental queries concerning IAEA publications.

Alternatively, contact IAEA Publishing:

Publishing Section
International Atomic Energy Agency
Vienna International Centre, PO Box 100, 1400 Vienna, Austria
Telephone: +43 1 2600 22529 or 22530
Email: sales.publications@iaea.org
www.iaea.org/publications

Priced and unpriced IAEA publications may be ordered directly from the IAEA.

ORDERING LOCALLY

Priced IAEA publications may be purchased from regional distributors and from major local booksellers.

Printed and bound by CPI Group (UK) Ltd, Croydon, CR0 4YY

06/07/2026

02160600-0012